HISTORICAL ROOTS OF COGNITIVE SCIENCE

AF785171

SYNTHESE LIBRARY

STUDIES IN EPISTEMOLOGY,

LOGIC, METHODOLOGY, AND PHILOSOPHY OF SCIENCE

Managing Editor:

JAAKKO HINTIKKA, *Florida State University, Tallahassee*

Editors:

DONALD DAVIDSON, *University of California, Berkeley*
GABRIËL NUCHELMANS, *University of Leyden*
WESLEY C. SALMON, *University of Pittsburgh*

VOLUME 208

THEO C. MEYERING
Department of Philosophy,
Leyden University, The Netherlands

HISTORICAL ROOTS OF COGNITIVE SCIENCE

The Rise of a Cognitive Theory of Perception from Antiquity to the Nineteenth Century

KLUWER ACADEMIC PUBLISHERS
DORDRECHT / BOSTON / LONDON

Library of Congress Cataloging in Publication Data

Meyering, Theo C.
Historical roots of cognitive science : the rise of a cognitive theory of perception from antiquity to the nineteenth century / Theo C. Meyering.
p. cm. -- (Synthese library ; 208)
Bibliography: p.
Includes index.
ISBN 0-7923-0349-0 (U.S.)
1. Visual perception--History. 2. Perception (Philosophy)--History. 3. Cognitive science--Philosophy--History. I. Title. II. Series.
BF241.M45 1989
153.7'09--dc20 89-15270

ISBN 0-7923-0349-0

Published by Kluwer Academic Publishers,
P.O. Box 17, 3300 AA Dordrecht, The Netherlands.

Kluwer Academic Publishers incorporates
the publishing programmes of
D. Reidel, Martinus Nijhoff, Dr W. Junk and MTP Press.

Sold and distributed in the U.S.A. and Canada
by Kluwer Academic Publishers,
101 Philip Drive, Norwell, MA 02061, U.S.A.

In all other countries, sold and distributed
by Kluwer Academic Publishers Group,
P.O. Box 322, 3300 AH Dordrecht, The Netherlands.

printed on acid free paper

x
Kart-Y 219

All Rights Reserved
© 1989 by Kluwer Academic Publishers
No part of the material protected by this copyright notice may be reproduced or
utilized in any form or by any means, electronic or mechanical,
including photocopying, recording or by any information storage and
retrieval system, without written permission from the copyright owner.

Printed in the Netherlands

"Als wesentlichsten Fortschritt der neueren Zeit glaube ich die Auflösung des Begriffs der Anschauung in die elementaren Vorgänge des Denkens betrachten zu müssen, die bei Kant noch fehlt, wodurch dann auch seine Auffassung der Axiome der Geometrie als transzendentale Sätze bedingt ist. Es sind hier namentlich die physiologischen Untersuchungen über die Sinneswahrnehmungen gewesen, welche uns an die letzten elementaren Vorgänge des Erkennens hingeführt haben, die noch nicht in Worte fassbar, der Philosophie unbekannt und unzugänglich bleiben mussten, so lange diese nur die in der Sprache ihren Ausdruck findenden Erkenntnisse untersuchte." (Hermann von Helmholtz, 'Die Tatsachen in der Wahrnehmung', 1878).

Daguerreotype showing Helmholtz as a young man at about the age of twenty, presumably around the time when he graduated from the medical school at Berlin in 1842, where his teacher Johannes Müller, the founding father of the rich research tradition in physiology in 19th century Germany, held the world's first 'Professorate in Physiology'.

Helmholtz shown again towards the end of his extraordinarily fertile and versatile career at the age of 60, this time as the renowned professor of physics, back again at the same Friedrich-Wilhelm Institute of Medicine and Surgery of the University of Berlin, where he had received his own education.

TABLE OF CONTENTS

PREFACE

Cognitive science, in Howard Gardner's words, has a relatively short history but a very long past. While its short history has been the subject of quite a few studies published in recent years, the current book focuses instead on its very long past. It explores the emergence of the conceptual framework that was necessary to make the rise of modern cognitive science possible in the first place.

Over the long course of the history of the theory of perception and of cognition, various conceptual breakthroughs can be discerned that have contributed significantly to the conception of the mind as a physical symbol system with intricate representational capacities and unimaginably rich computational resources. In historical retrospect such conceptual transitions—seemingly sudden and unannounced—are typically foreshadowed in the course of enduring research programs that serve as slowly developing theoretical constraint structures gradually narrowing down the apparent solution space for the scientific problems at hand. Ultimately the fundamental problem is either resolved to the satisfaction of the majority of researchers in the area of investigation, or else—and much more commonly—one or more of the major theoretical constraints is abandoned or radically modified, giving way to entirely new theoretical vistas.

In the history of the theory of perception this process can be witnessed at various important junctures. In the first part of this book I have focused, in particular, on the Aristotelian identity theory of perception; on the Alhazenian synthesis in optical theory during the Arab and European Middle Ages; on the radical impact of seventeenth century mechanicism and the attendant dissociation of a representational psychology of visual perception from physiological and mathematical optics; on the rise, and the vicissitudes, of a Cartesian inspired computational theory of mind; on the rise of rival empiricist learning theories of perception during the 17th and 18th centuries; on the persistent philosophical bias in rationalist and empiricist circles alike which served to identify cognitive activity with conscious activity, thus hampering the development of a full-fledged empirical and experimental psychology; and, finally, on the rise of a truly information-theoretical conception of the mind in the seminal work of the nineteenth century mathematician, physicist, and (neuro-)physiologist Hermann von Helmholtz.

From this perspective it transpires that the history of the theory of perception is characterized by the gradual emergence of a cognitive theory of perception according to which perception involves information processing of an essentially interpretive character. The implied radically novel conception of the human mind on the one hand helped to define an entirely new research program in cognitive psychology, whose impact has become ever more keenly felt especially after its relatively recent coalescence with various breeds of computational theories of mind in computer science, psycholinguistics, neuroscience, and the philosophy of mind. On the other hand, it immediately raised acute epistemological problems as well. For given this theoretical perspective, perceptual knowledge is seen as essentially based upon *insufficient evidence*. Helmholtz was one of the first philosopher-scientists to perceive the epistemological problem in such terms, to pose the central question of how to reconcile a truly information-theoretical account of perception with a theory of objective perceptual truth, and to initiate a novel research program that would represent an important contribution to the solution of that query.

Helmholtz embodied the ideal of the *homo universalis*. An extraordinary polymath, his accomplishments ranged over many fields of study, from measurement theory to theoretical physics, from physiological optics to metamathematics, from epistemology to cognitive psychology. He thus worked ('unorthodoxly') on both sides of the fence which later (neo-kantian, logical positivist and analytic) philosophers deemed necessary to erect between philosophy and empirical science. Thus various promising Helmholtzian ideas were *a priori* ruled out of court by subsequent generations as 'rife with conceptual confusions'. This holds particularly true of Helmholtz's pioneering theory of unconscious inferences, which has been of such foundational importance for the conceptual framework of cognitive psychology as well as for that of cognitive science in general, but which even now tends to be passed off by some contemporary philosophers as 'a mythology of mental processes' and again—a very favorite verdict—as 'a tangle of conceptual confusions'.[1] It will be the main concern of the second part of this book to describe and evaluate Helmholtz's naturalism and his lasting contributions to both epistemology and cognitive psychology, as well as to discuss the relevant theoretical controversies in nineteenth century physiological optics Helmholtz was engaged in as a consequence of his philosophical point of view.

1 P.M.S. Hacker, 'Helmholtz's Theory of Perception: an Investigation into its Conceptual Framework', *forthcoming*.

Yet this study aspires to offer more than unadulterated intellectual historiography. In addition, I also hope to adduce positive arguments on behalf of the philosophical thesis inherent in naturalistic epistemology. The Kantian demarcation within the realm of legitimate knowledge between philosophy and empirical science is bound to prove detrimental to both, if it is taken to imply that philosophy is an independent discipline entitled to some ultimate verdict on the study of the structures of man's cognitive and perceptual faculties in virtue of the claim that philosophy alone enjoys privileged access to some special set of strictly *a priori* insights. To be sure, the distribution of intellectual tasks prescribed by Kant failed to prevent the rise of empirical psychology and of psychophysiology in the course of the nineteenth century. Nevertheless, the Kantian proclamation still reigns supreme among quite a few contemporary philosophers—albeit now in the guise of some updated version of analytic philosophy. Indeed, up to this very day many philosophers still adhere to the view that the mind-body problem is a strictly *conceptual* issue which can *only* be illuminated and resolved by philosophical analysis of the relevant concepts in ordinary discourse. The notion that ordinary language might itself contain a hidden, but all the more tenacious *theory* about the etiology of human behavior, and that therefore the apparently 'intuitive evidence' our linguistic habits seem to offer may at best *compete* in the arena of philosophical argument as a (relatively naïve) *rival interpretation*, but certainly not as the highest *arbiter* in matters psycho-philosophical, is an insight that has not yet dawned, it seems, upon the philosophical orthodoxy at large.

In contrast, the current study emphatically favors the non-Kantian, naturalistic view that philosophical and psychological questions are theoretically continuous with each other. They are, that is, mutually relevant and can only be resolved by a fruitful interaction between the respective disciplines. Consequently, it is to be expected that the rise and the radical development of experimental studies of cognitive faculties and their neurological substrata over the past one hundred years and especially during the post-war period, implicitly comprises quite specific consequences for the sophistication, adjustment, or rejection of current philosophies of mind. Conversely, philosophical considerations can in principle be expected to yield fruitful theoretical suggestions for the interpretation of experimental psychological research. Thus in modern cognitive science philosophical and experimental research tend to go hand in hand. In fact, cognitive science can be defined as a multidisciplinary effort with strong empirical overtones attempting to solve long-standing theoretical problems in the philosophy of mind and the theory of knowledge by building plausible information-processing models of how the mind 'ticks'. These models are subject to constraint structures in which elements of the *entire* interdisciplinary field, whether conceptual, theoretical, or empirical, may serve as determining vectors depending on the degree of their presumed epistemological stability. Consequently, philosophical questions concerning the human mind can no longer be decided 'internally', on no other grounds than purely *a priori* considerations and with a total neglect of the results of relevant research within experimental psychology. On the contrary, just because philosophical presuppositions do play a role, explicitly or otherwise, in the determination of rival constraint structures in theoretical psychology, whose relative fertility must be proven in ongoing research, these basic philosophical assumptions, just as all other theoretical assumptions in empirical science, become accountable to empirical evidence as well.

In fact, the rational reconstruction of the history of the theory of perception as pursued in this book, at the same time itself represents an exercise in naturalistic epistemology. For the possibility of reconstructing—that is, of making true sense of—the histories of 'philosophical' epistemology and the 'empirical' theory of perception *in isolation from each other* is expressly denied. Rather it is shown that philosophical and empirical questions are directly and inextricably intertwined. Accordingly it is argued that epistemological questions can only be rendered intelligible at any given time relative to a comprehensive theoretical enterprise in which philosophical and empirical developments interact and are jointly relevant. The admittedly ambitious project embarked upon in the present book thus also expands the naturalistic thesis not by demonstrating its truth, but by showing its success if adopted as a research program for philosophico-historical inquiry. This in turn generates positive arguments on behalf of naturalism which instead of being purportedly derived from evident truths or *a priori* insights (a pitfall to be avoided at all cost as it would undermine the very proposition thus defended), would be derived solely from its illuminating force and its practical effectiveness as an instrument of historical analysis. Again, it is a plausible corollary of naturalistic epistemology that the history of so-called philosophical epistemology should itself be susceptible to successful analysis by means of historiographical models specifically designed in order to describe, or to account for, the development of *theoretical science*. A positive test of this consequence should thus count as a corroboration of the theoretical proposal inherent in epistemological naturalism.

Too much time has elapsed since the ideas expressed in this book first began to take shape, for me to distinctly remember the rather variegated philosophical influences that have helped to determine the eventual outcome of my work. I still feel immensely grateful to have benefited for so many years from the American academic scene in general, and in particular from the very diverse intellectual attractions the Berkeley campus had to offer during the years of my graduate training there. Those were years of great intellectual excitement, due in the first place to the quality and the diversity of the philosophical Faculty at the time, and secondly to the flux of prominent scholars in various fields of study visiting this intellectual Mecca on the West Coast (and here I am not just referring to the likes of Professor Philip Swallow, immortalized by David Lodge). Thus I remember—not necessarily in order of vividness—the vigorous courses taught by John Searle; or Barry Stroud's exercises in historical and conceptual analysis; or the subtlety of Benson Mates, who combined dignity with an irrepressible sense of humor; Paul Grice's power of analytic thought; Charles Chihara's pungent style of argument; or Carl Hempel's acumen, open-mindedness and irresistible narrative charm—to mention but a few of the many precious recollections I still cherish of the intellectually stimulating ambiance Berkeley provided me during my stay.

However, my greatest intellectual debt, especially with regard to the present book, I owe without doubt to two of my former teachers of philosophy, Hans Sluga and Paul Feyerabend. Their appreciation, each in his own very distinct way, of the historical, cultural, and theoretical embeddings of philosophical enquiry, and indeed of the contextual nature of virtually all intellectual pursuit worthy of the name, has had a lasting influence on my own naturalistic inclinations 'in matters epistemological'. Feyerabend's seminars, conducted in a fashion which, to put it mildly, deviated considerably from conventional styles of formal instruction, always seemed to hold the promise of something like a true intellectual happening, breathing an air of novelty and of stunning surprise. A colorful debater, with a malicious style, taking unexpected turns and bewildering his audience with his imaginative ideas, he inspired as much as he provoked. To more rigorous minds perhaps no more than a frivolous source of countersuggestion, I am convinced that many of Feyerabend's critical ideas will stand as an enduring monument of mid-twentieth century philosophical innovation.
Feyerabend, of course, needs no praise. Nor is he very likely to relish it. For the anarchist, epistemological or otherwise, when eulogized with reverence and acclaim, usually suspects the impending advent of his gravest foe, general respectability. No wonder then, that the old master's scorn has always been more readily incited by applause than by criticism. So be it.

No less beneficial than Feyerabend's impact has been the influence of Hans Sluga's philosophical ideas. His critical assessment of analytic philosophy (right in 'The Bear's Lair', so to speak) generated new insights and methods of philosophical understanding that were a welcome supplement to the prevailing methods of formal and conceptual analysis with which I had become abundantly acquainted during my former philosophical training (and which I still hold in due regard). While Feyerabend had emphasized that *science* can only be understood as a historical phenomenon, Sluga extended this insight to the nature of *philosophical* theory formation as well. Sluga's own scholarly work on Frege provides an excellent example of the wide range of new and valuable insights that can be gained once this post-analytical conception of philosophy is put to work in the concrete context of actual philosophico-historical enquiry. Clearly, his basic point of view has helped to shape the methodology of the present book and, in general, has left indelible marks on my own appreciation of the relation between philosophy and theoretical science, and of the ineradicably historical nature of both. In addition, I am also deeply grateful for the encouragement I received from Hans Sluga, who indefatigably spurred me on to go ahead and publish this book.

Furthermore, from among the many scholars referred to in the footnotes I owe a special debt of gratitude to the pioneering research of A.C. Crombie and to the meticulous historical studies of David C. Lindberg. Their extensive work, especially in the area of medieval and Renaissance theories of vision, has proved an invaluable asset to my own attempts at historical reconstruction.

I also wish to thank Richard Rorty for his insightful philosophical remarks and his valuable editorial comments, both of which I have tried to incorporate as well as I could.

Many friends and colleagues, in one stage or another, have obliged me with their encouragement, their helpful comments or their critical suggestions. In particular I wish to thank Diedel Kornet, Herman Philipse, and Françoise Wemelfelder of Leiden University; Gerard de Vries of the University of Limburg; Colin Brown of the Max Planck Institute for Psycholinguistics; Paul van Seters of Tilburg University; and my old Berkeley friend Martin van den Toorn.

Last but not least I owe very special thanks to my friend Adri for her patience and her support.

Peter Hock of Leiden University provided me with graphical assistance for some of the illustrations that proved beyond my own (computer aided) design skills.

Publication of this book has been made possible in part by a grant from the Netherlands Organization for Scientific Research (N.W.O.), which I gratefully acknowledge.

Theo C. Meyering
Amsterdam, 1989

I

INTRODUCTION

First philosophy as the foundation of all truth and wisdom must be epistemology. This is a notion so widely received that the philosopher of today tends to accept it as the common wisdom of his trade. For, surely, no proposition is admissible as knowledge unless rigorously tested. And it goes without saying that the branch of philosophy establishing these tests, thus formulating the criteria of knowledge, must, on pain of vicious circularity, be prior to, and proceed independently of, any material cognitive claims. First philosophy, therefore, provides no substantial knowledge. It is a purely formal discipline, necessarily critical and epistemological.

However, it would not be the first time that propositions seemingly necessary and general turned out to have a history and to be disputable after all. The above argument, too, however seemingly ineluctable, is neither of all times nor did it remain unchallenged since its incorporation among the wisdoms of the West. As for the first point: it was only since Descartes that the 'motherhood' of philosophy was transferred from system-minded wonder to systematic doubt; from a study in search of the essences of *things* to one that takes *human cognition* as its point of departure. Formerly, first philosophy comprised ontology and metaphysics, inquiries into the basic structure of reality. Epistemology was a special application of the insights thus acquired. Aristotelian epistemology was conceived in terms of the *antecedently available* vocabulary of the act-potency analysis which itself resulted from the fundamental question of Aristotelian metaphysics concerning the possibility of change. Here theory of knowledge was not yet a formal science claiming to *justify* the products of knowledge. Rather, it was a kind of comprehensive 'epistemics' *describing* cognitive processes, and explaining them in terms of categories which also rendered intelligible the operation of other natural processes.

It was only in the hands of Descartes that first philosophy became epistemology: inquiry into the (empirical and conceptual) conditions of human experience. This revolution in the history of philosophy was a radical and universal one. The fundamental question of practically all post-

Cartesian philosophy ceased to be the question of the 'things themselves'. Instead, the scope and limits of human knowledge, or the 'mental conditions of experience' were taken as the ultimate point of departure of all philosophical inquiry.[1] The principal reason for this transition must be sought in the collapse of Aristotelian physics and metaphysics and their replacement by a deistic theology and a mechanistic world view. Henceforth the only remaining reality was the reality of matter and its unbreakable natural laws. No 'deeper' metaphysical reality existed as final cause of all being and knowing. The book *Genesis* was meant for the simple-minded. For theoretical purposes, however, God was best regarded as a *deus otiosus*.[2] Similarly, no metaphysical or theological warrant any longer existed for the validity of human knowledge.[3] Thus the ontological ground which in ancient philosophy declared knower and known to be one in the act of knowing dropped out.

For the Greeks the source of the intelligibility of the phenomena was not located in the human mind but in separate metaphysical entities (the ideas of Plato, or Aristotle's self-absorbed God). Thus the realm of (material) phenomena, too, existed objectively, independently of the perceiving mind. The phenomena were a formal product not of the synthesizing activity of the knowing subject but of the formative effects of non-worldly (formal or final) causes. However, given that the formal principle of knowledge is sought *outside* human consciousness and placed, instead, in an immaterial reality *inaccessible to the senses* neither psychology (or the doctrine of cognitive consciousness) nor physics (or the doctrine of material phenomena) can

1 Cf. Kenneth Dorter, 'First Philosophy: Metaphysics or Epistemology?', *Dialogue* 11, 1 (Mar. 1972), p.3.

2 Cf. Descartes' scheme of the creation of the universe in *Le monde*, which he withheld from publication when he learnt about Galilei's condemnation.

3 Note that the proposition of God's existence is presented as a *theorem* within the Cartesian theory of knowledge, not as an unproven *premise* supported by faith and revelation. This, of course is a highly contentious issue which up to the present day has been the subject of animated discussion directed at the so-called problem of the Cartesian circle. Yet, whatever the merits of the Cartesian argument, it is clear that the edifice of knowledge in Descartes' view is built upon the Cogito and the powers of a self-reflective mind, that is, on critical epistemology rather than on inspired theology. For the philosophical problems surrounding the Cartesian circle, cf. Alan Gewirtz, 'The Cartesian Circle', *Philosophical Review* L, 4 (July, 1941) pp. 368-95; id., 'The Cartesian Circle Reconsidered', *Journal of Philosophy* 67, 19 (Oct. 8, 1970) pp. 668-85; Anthony Kenny, 'The Cartesian Circle and the Eternal Truths', ibid., pp. 685-700; Harry G. Frankfurt, *Demons, Dreamers and Madmen: The Defense of Reason in Descartes' Meditations*, (Indianapolis, 1970), esp. ch. 15.

reasonably serve as an adequate base for philosophical reconstruction. For then the formal identity of knower and known which is thought to condition the very possibility of knowledge as such can be derived only from ontological (metaphysical) principles.

Since Descartes, however, the *Cogito* of individual consciousness becomes the pivot of the (only knowable) universe. The 'things themselves' are no longer immediately accessible. Henceforth knowledge must found itself. Moreover, in order to penetrate into the essence of things common sense alone no longer suffices: it must also be applied correctly.[1] For now things are 'hidden' rather than manifest.[2]

Thus, more acutely than ever before the need is felt to give a *methodological* account of all knowledge. For the cognitive process is now regarded simply as the mechanical (i.e., in Descartes' view, the only intelligible) transfer of physiologically transformed information without there being any metaphysical ground for believing that the essential attributes of things will be retained in the subjective sensations (the former phantasms of sense[3]). The question whether reality really is what it appears to be now becomes radically problematic. Since phenomenal experience distorts not only occasionally but systematically, it is best regarded as a veil which hides reality rather than revealing it. Thus the mind, not the eye, has the sole power of discernment: it alone truly discriminates between real and merely apparent 'information'.[4] However, this implies that the mind in its turn may *impose* its structures upon reality as perceived. Consequently, the need arises of an objective justification of all cognitive claims. And this justification is sought—can be sought nowhere else, it seems—in a comprehensive theory of human consciousness and of the perceptual process which will bridge the gap between knowing and being. It is this (Cartesian) research project modern

---

1 *Discourse on Method*, E.S. Haldane and G.R.T. Ross (trs.), *The Philosophical Works of Descartes*, 2 vols., (Cambridge, 1911-1912; corrected ed., 1931; paperback ed. New York, 1973), I, pp. 81-2. This edition will be referred to below as HR.

2 Only propaganda on behalf of the new science, and of the new common sense that came along with it, would present the objects and their real natures as manifest, manifest to reason, for example, or manifest to the expert practitioner of mathematically interpretable experiments. The latest victim of this propaganda seems to be Sir Karl Popper who takes the era since the Scientific Revolution to be decisively characterized by the doctrine that *truth is manifest* [cf. K. Popper, *Conjectures and Refutations*, (London 1963) p. 5].

3 Thomas of Aquinas, *Summa theologica*, Q 85, Art. 1

4 Cf. Descartes' example of the melting piece of wax in *Meditation* II, HR I 154-5

philosophers—the 'empiricists' as well as the 'rationalists'—will be thoroughly engaged in for at least two centuries to come.

In a sense Kant's later contribution was no more than a radicalization of this project. By eliminating all contingent quasi-empirical and psychological elements he made the question of justification strictly dependent upon a transcendental (nonempirical) analysis of the phenomenon of human knowledge. But the interpretation of the distinction between appearance and reality as an *epistemological* distinction between mental phenomena and a mind-independent reality is not Kantian in origin but Cartesian. This conception contrasts sharply with the classical interpretation of the distinction between phenomenon and reality as an *ontological* distinction between material appearances on the one hand and the reality of metaphysical entities not accessible to physical inquiry on the other.[1]

An interesting consequence of the new (Cartesian) view was the extension it implied of the legitimate claims of science. For now the phenomena were no more than the 'front side' (the side facing experience, as it were) of the things themselves. Therefore, *to investigate these phenomena was to investigate all there was*. Moreover, provided research was carefully guided by the canons of a sound methodology (based upon an exhaustive and comprehensive theory of cognitive consciousness) the scientist was in principle capable of decoding the systematic distortion inherent in ordinary as well as scientific experience, thereby obtaining an indirect view of the things themselves, the true nature of reality. Thus, natural science in the broadest sense of the word, provided it was founded upon a radical theory of knowledge and a strict methodology (Descartes) or else complemented with a scientific psychology (British empiricists and French sensationalists), constituted the whole of the true universal science.

This claim of the cognitive universality of modern science constituted a radically new phenomenon in the history of ideas. It brought in its wake a fundamental change in the traditional distribution of intellectual tasks. Until then the powerful Thomist synthesis of the 13th century had forged rational thought and the *Sacra doctrina* into one all-comprising rational-spiritual system whose beneficent unity was to be intellectually gratifying as well as stimulating for generations of scholars to come, theologians and philosophers alike. Within this Aristotelian-Thomistic system reason and revelation could not possibly clash. On the contrary, they were seen as converging and *complementary* sources of true knowledge. Thus, not only

1 Cf. Kenneth Dorter (1972), p. 17.

was theology (or the knowledge of faith) intellectualized, but, conversely, the motive for the free cultivation of natural science was spiritualized. Its aim was not merely to master nature nor merely to describe it. Rather, it aimed at contemplating the cosmos and understanding the mysteries of the created universe, with deepened knowledge of God as its highest ideal.

For more than three centuries this grand intellectual structure stood its ground, internal friction and Ockhamist opposition notwithstanding. During the 17th century, however, its gradual decline came to a final collapse. And the new intellectual order arising in its place looks as though it is the mirror-image of the old one. The traditional *ancilla-domina* relation of science and philosophy with respect to theology virtually turns into its very opposite. The plea between faith and reason—uncomfortable bedfellows anyway, Thomas's genius notwithstanding—is definitely settled, it seems, in favor of the latter.

But the distinction between empirical science and philosophy, too, gets more sharply articulated. On the one hand, the need of gaining insight into the true foundations of the phenomena is no longer satisfied by (Aristotelian) metaphysics. On the contrary, modern physics claims *exclusive* rights in this respect. On the other hand, science no longer counts the complex processes of sensory perception as belonging to its proper field of inquiry. Thus, optics is no longer a comprehensive theory of *vision* which investigates the entire phenomenon of light in all its aspects (both as *lux* and as *lumen*[1]).

---

1 The Scholastics regarded *lumen* as the species of *lux*, that is, as the representation of *lux* in the medium by means of which *lux* is seen. By contrast, *lux* is a fixed property of a luminous body. According to Jean Buridan (fl. ca. 1330), in order that color is seen *lumen* and the species of color must act together: "lumen and (the species of) color are sufficient for affecting vision and impressing in it species sufficient for vision". (*Questiones et decisiones physicales insignium virorum*, George Lockert (imp.), (Paris, 1518), fol. 14va; quoted in David C. Lindberg, *Theories of Vision from al-Kindi to Kepler*, (Chicago, 1976, p. 133). He goes on to argue, rather crudely, that as perception does not occur through the reception of the object itself it is the reception of its species in sense or the organ of sense which constitutes perception (ibid., fol. 19rb).

Clearly, in the Scholastic account of vision the 'subjective' and the 'objective' manifestations of light are not at all distinguished *as such* either ontologically or methodologically. Vasco Ronchi has drawn attention to this aspect of medieval theories of vision and to its contrast with modern conceptions (cf. his *Optics, The Science of Vision*, (NY University press, 1957), esp. ch. I.). However, Ronchi's account of the ancient and Scholastic concepts of *lux, lumen, species* and color and their theoretical and philosophical underpinnings seems a little confused at times (cf. esp. pp. 14-5, 17-9, 32). For an incisive criticism of a central doctrine of Ronchi's work, cf. David C. Lindberg and Nicholas H. Steneck, 'The Sense of Vision and the Origins of Modern Science', in *Science,*

Rather, it becomes exclusively geometrical or physical optics, a study of objective light alone. Questions concerning visual perception ('subjective light') it now recognizes as *psychological* questions concerning conscious cognitive effects of mechanical activity. Such questions belong to, or are now assigned to, the proper area of philosophy and epistemology. Thus the various natural sciences become increasingly autonomous while they organize themselves methodologically and institutionally into separate research institutions.[1]

By contrast, modern philosophy is left in particular with the difficult task of accounting for man's perceptual and cognitive capabilities. Paradoxically enough, the very astonishing insights of modern mathematical physics the human mind has shown itself capable of comprised a mechanistic philosophy of nature which rendered that same intellect—its manner of operation as well as its relation to the external world—more obscure and unintelligible than ever before. Thus seventeenth century man lost in knowledge of self what he gained in knowledge of nature. For whereas for the natural sciences the mechanistic hypothesis defined a research program which proved to be immensely fruitful, for philosophy and psychology it seemed to imply the opposite: an all but impossible task. Indeed, in a whirling universe of central point forces and colliding fragments of inert dead matter the human mind is little more than a receding shadow, an intangible delusion. Nothing appeared more tempting than to subsume the laws of thought and consciousness under the universal laws which proved to govern *all* motion. And yet, at the same time, nothing appeared more impossible than just that: to reduce mental to physical phenomena, or to derive mind from mere matter. Thus, with razor-sharp clarity mechanicism carved out the psychophysical problem which, ineluctably, would remain a, or perhaps *the*, central problem of philosophy up to the present day.

Now in order to show how this complex problem area led to fundamental changes in the overall theoretical framework of modern versus medieval and classical theories of perception I will first try to isolate the basic philosophical ideas or 'negative heuristics'[2] underlying the various medieval

---

Medicine and Society in the Renaissance, Essays to Honor Walter Pagel, Allen G. Debus (ed.), (New York, 1972).

1 Cf. Thomas S. Kuhn, 'Mathematical versus Experimental Traditions in the Development of Physical Science', *The Journal of Interdisciplinary History*, 7 (1976), pp. 1-31.

2 Cf. I. Lakatos, 'Falsification and the Methodology of Scientific Research Programmes', in *Criticism and the Growth of Knowledge*, I. Lakatos and A. Musgrave (eds.), (Cambridge 1970), pp. 91-196.

theories of perception which guided their development and constituted their common core notwithstanding their great theoretical divergences in other respects (see chapters II and III). As my primary concern will be to bring out the *epistemological* contrast with the subsequent seventeenth century program, all sorts of (optical) details will be left undiscussed until in chapter IV I will extensively deal with the important role played by the 'hard core' of medieval theories of perception in shaping the various optical traditions of the time. This will make clear, I hope, that a profound revolution in basic ideas and philosophical presuppositions was necessary for the achievement of a long overdue break-through in optical theory (including the theory of vision).

A third theme, which is important and intriguing as such, can only be briefly touched upon within the framework of the present book (see chapter V). It involves a tentative hypothesis which in the course of my inquiry suggested itself with increasing force. Put briefly, the suggestion boils down to the idea that ontologically as well as methodologically the true onset of the momentous intellectual shift that would ultimately mathematize physics and mechanize the overall world view is to be found in the development of medieval *optics* rather than in that of mechanics. For it was in optics that theoretical and experimental perspectives had traditionally been very close. Moreover, it was in the development of optical theory that the mathematical approach had increasingly shown itself capable of explaining *physical* details as well. Finally, it was optics which had been closely associated of old with certain cosmological conceptions derived from a magico-mathematical tradition, conceptions which easily lent themselves to re-interpretation along mechanistic lines.

Having thus dealt with the profound conceptual changes that marked the transition from medieval optical theory to the 17th century theory of (visual) perception I will continue my naturalistic reconstruction of the history of (optical) epistemology by discussing at length the impact of mechanicism upon the theory of (perceptual) knowledge (see chapter VI). While the former success of Scholastic philosophy had made it the natural source of inspiration for empirical theories of vision, the tables were now turned. The triumph of the mechanicist doctrine implicit in Keplerian dioptrics and shortly to be elaborated philosophically by Descartes now exposed the radical aporia of philosophical epistemology. Now was the time for philosophy to follow the lead of science (this explains Locke's explicit underlabourer conception of philosophy). Now it was philosophy's task to keep in pace as well as in tune with the triumphant march forward of empirical science. The latter dictated the certitudes of a new metaphysics, a corpuscular philosophy and a mechanicist doctrine. But precisely these certitudes created a radical crisis in traditional philosophy. The new task was to re-establish the natural link between knowing and being so radically severed by the new science.

I will focus upon two related developments in the emerging representationist research program into the theory of (perceptual) knowledge. One is the rise of an *information theory of perception*. The other phenomenon is the rise of *justificationist methodologies*.

Thus, on the one hand a new notion of consciousness seemed to be forced upon philosophy by the mechanicist doctrine. But the conception of mind as an active processing device was too radical to be adopted at once. Moreover, if adopted, it would seem hopelessly to endanger what epistemologists were at pains to rescue: the objective validity of (some) perceptual knowledge. Thus Descartes' optical epistemology required a dualism of judging and sensing. And when he discussed the means by which we perceive distance he developed neither a learning theory nor an information-theoretic account of the processing of subconscious data. He rather invoked, in Keplerian fashion, a "natural geometry" establishing necessary links between sensory stimuli and perceptual responses.

Malebranche, by contrast, was more keen on describing quasi-intellectual processes dealing with subconscious data. But he developed no learning theory either. Instead he collapsed these processes and their results into mere sensation, ascribing them to the intervention of God and showing their liability to error, thus underscoring his favorite philosophical thesis that reliable knowledge is to be obtained through thought and intuition, while the senses generally deceive.

On the other hand, the empiricists did develop learning theories but these tended to be purely mechanical (thus Hume's analogy between the principles of gravitation and of association) and they were allowed to act on *conscious* data alone (thus Berkeley's principle that an idea not itself perceived cannot be the means for the perception of another idea).

I will trace the various theoretical developments into the 18th century, showing how the representationist research program gradually crumbled as a consequence of the internal tensions between its rich positive heuristic and its restrictive negative heuristic. I will argue that the 'missing link' is prepared, finally, in the tradition of early German Romanticism which inspired the profound conceptual breakthrough achieved in the 19th century research program initiated by Hermann von Helmholtz. Here, at last, a truly information-theoretical account of human perception came to full fruition, implying on the psychological side a theory of subliminal cognitive activity, and on the epistemological side a theory of hypothetical realism.

Now with regard to the other aspect of the representationist research program mentioned above, viz., the rising need of justificationist methodologies, another insight can be derived from my naturalistic inquiry. The need for justification, as well as the distinction between contexts of discovery and contexts of justification, seem to express eternal truths. Thus not only is *justification* commonly expected to proceed on the basis of a priori insights. But the *need* for justification, too, is presented as self-evident and as independent of any theoretical views one might hold. Thus, the modern conception that at the basis of all knowledge lies a theory of knowledge or that first philosophy is necessarily epistemology—a notion generally adopted since Descartes, as I have pointed out—has become the irresistible common sense of contemporary philosophy. However, if one studies the history of ideas a little bit more closely, it becomes quite clear that the need for *standards* arises only when fundamental *views* seem to falter, especially when those views (as was the case with the Aristotelian system of the Scholastic era) are so universal and all-embracing that any epistemology can only be formulated in terms of the basic metaphysical framework. Thus, Aristotelian physics was naturally embedded within the fundamental metaphysical theory of change of which, one might say, it was no more than a special application (change from ignorance to knowledge).

Only when this comprehensive theory broke down (and what I will call the *identity theory of perception* along with it) the (Cartesian) need to justify knowledge arose. This need, moreover, was exacerbated by the specific *world-view* suggested by modern science and universally adopted, which explained natural reality in terms which had no ostensible connection with the concepts of phenomenal experience. To be sure, the possibility of illusion was known since antiquity and classical cases were discussed during the Middle Ages in circles of so-called academic skeptics. But the new philosophy raised the specter of wholesale deception which only a sound psychological theory of cognitive awareness, of mind and body, of the relation between primitive data of consciousness and peripheral stimuli, of the origin and development of concepts, etc., could conceivably overcome. Paradoxically enough, the scientific revolution of the 17th century has thus led to the profound insight that (*pace* Popper[1]) *truth is not manifest*. Thus

1 Cf. K. Popper, *Conjectures and Refutations*, (London 1963), p. 14, n. 2.

science is henceforth regarded as in need of an explicit method[1] and of an *epistemology* which justifies its results.

Its primary task is (or so it was conceived to be) to formulate a theory of mind *compatible with the results of mechanical physics and geometrical and physiological optics*. Within this boundary condition it had to develop an adequate explanation of the discrepancies *established by science* between phenomenal appearance and reality as determined by intellectual or experimental means.

The theory of ideas and the representative theory became the dominant programs of epistemology. Stimulus and experience were no longer supposed to be formally *identical* (as the Scholastics thought) but one hoped to re-establish the link between the *ordo essendi* and the *ordo cognoscendi* by assuming instead some kind of *resemblance* between phenomenal and stimulus properties. This solution of the problem of perception, however, turned out to raise more questions than it answered. And it was these questions which determined the internal dynamics of the respective research programs well into the 19th century.

At this point we have reached the epistemological problem facing the 19th century and constituting the (latent) background of various theoretical controversies in physiological optics at the time (see chapter VII). And it is to the solution of this problem to which Helmholtz's epistemology as well as his seemingly disparate scientific research addresses itself. Thus, pursuing lines of development in the theory of knowledge, our naturalistic inquiry again leads up to Helmholtz's comprehensive program into the foundations and the scope of human knowledge.

Accordingly, the history of the theory of perception might be schematically summarized as follows (cf. schema on page 11). Whereas the Peripatetics regarded perception as involving literal *in-form-ation* and perceptual knowledge as based upon the absorption of the true objects of perception, the (post-) Cartesian theoreticians realized that perception involves information processing and 'decoding' of sensory inputs. Thus, perception (according to them) stands to its objects as hypotheses stand to the available evidence. However, they still regarded perceptual knowledge as based upon *sufficient evidence*. By contrast, modern theoreticians have come to realize that even this latter remnant of naïve realism is untenable. According to them perception involves information processing of an essentially interpretive

1 Cf. Descartes, HR I 81-2.

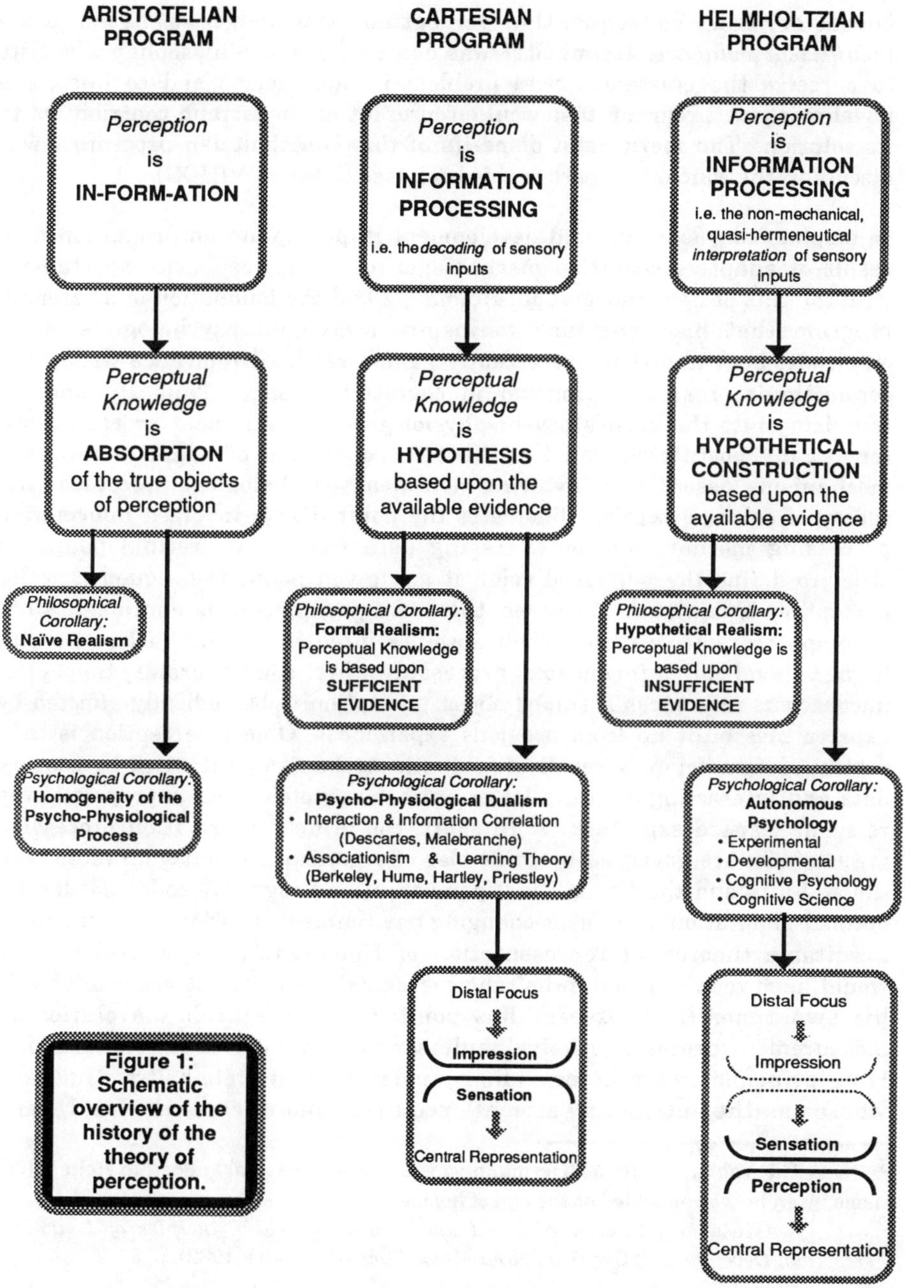

Figure 1: Schematic overview of the history of the theory of perception.

character and, consequently, perceptual knowledge is based upon *insufficient evidence*. Helmholtz[1] was one of the first philosopher-scientists to perceive the epistemological problem in such terms and to initiate a novel research program that would represent an important contribution to its solution. The merits and demerits of this Helmholtzian program I will discuss in the concluding part of this book (see chapters VIII-XI).

In emphasizing learning and development in perception and cognition as a result of adaptive cognitive mechanisms matching feedback expectancies with the flux of new information, Helmholtz laid the foundation of a research program that has continued to inspire individual psychologists of an empiricist bent and that has recently gained renewed relevance within the connectionist research program in cognitive science. Thus Helmholtz's theorizing into the brain's psycho-physiological measurement processes systematically postulated and identified the operation of adaptive feedback mechanisms essentially involving nonlinear psychological processes. His theory of color perception illustrates the point. For it involved information processing mechanisms for averaging data from many retinal points in order to define the perceived color at any given point. Consequently, color perception is nonlocal. Moreover, the averaging process is nonlinear, since it computes relative rather than absolute intensities.[2] Similarly, at a still higher level of information processing, his controversial theory of unconscious inferences assumed object perception to be radically affected by expectancies built up from previous experiences. Object perception is thus portrayed as a complex result of feedback mechanisms activated by sensory data and interacting with the latter until they match in an appropriate way relevant learned expectancies. In short, the brain's information processing mechanisms are governed by principles of organization that guarantee both the stability and the flexibility of perceptual and cognitive codes needed for optimal adaptation to an ever changing environment. However, in both cases a suitable theoretical representation of Helmholtz's conjectural scheme would have required conceptual and mathematical tools not yet available in his own time. As Grossberg has pointed out, the brain's evolutionary measurement processes involve nonlinear systems, whose laws and underlying principles were at best dimly understood in Helmholtz's time and whose mathematical treatment required nonlinear, nonlocal, and

---

1 The 'von' in the name 'Hermann von Helmholtz' was only added to Helmholtz's name when he was ennobled at the end of his life.

2 S. Grossberg, *Studies of Mind and Brain, Neural Principles of Learning, Perception, Development, Cognition, and Motor Control*, (Reidel, 1982), p. 3.

nonstationary tools and concepts not yet available to him.[1] It is not until fairly recently that the relevant mathematical techniques and concepts have begun to be fully employed and implemented in experimental settings. Indeed, the full utilization of these mathematical and conceptual tools constitutes one of the defining traits characterizing the latest and certainly most promising research program that has recently emerged in the field of cognitive science, a program generally known as connectionism. But it is good to bear in mind that the theoretical inspiration defining the philosophical hard core of this program certainly originates with Helmholtz's pioneering ideas.

1 Ibid., p. 634.

II

RECONSTRUCTION OF THE HISTORY OF MEDIEVAL AND (POST-) CARTESIAN THEORIES OF PERCEPTION IN TERMS OF THE NEGATIVE HEURISTICS OF THEIR RESPECTIVE RESEARCH PROGRAMS. BASIC EPISTEMOLOGICAL CONTRASTS.

The central concern of medieval epistemologists in so far as they dealt with problems of perception was to construct an adequate theory of vision. And the overriding problem within the theory of vision was to *explain* (rather than to justify[1]) the transmission of information from the seen object to the seeing eye. That is, if the full scope of the problem of perception is seen, in modern terms, to stretch all the way from distal and proximal to peripheral and central variables, then what the Arab and medieval scholars found particularly intriguing and problematic was the link between distal focus and peripheral response. The need for distinguishing a proximal variable in perception was not felt. Nor, consequently, were questions raised concerning the 'ecological validity' or the potential cue value of the proximal characteristics relative to the distal events. Furthermore, the intraorganismic processing of the given in perception was not felt to be of crucial relevance to the theory of *perceptual* knowledge. To be sure, according to the Scholastics the sensible imprints of nature upon our senses are further refined by operations of abstraction whereby sensible forms (*species sensibiles*) give rise to intelligible ones (*species intelligibiles*)—which explains, or attempts to explain, how we arrive at universals and how they are related to sense experience. But—and this is a crucial difference with later epistemological theories—*this was not a process of cue utilization*, i.e., of finding out the real object of perception by means of a meager set of data which must be held at abeyance and double-checked. On the contrary. According to the Aristotelians, at any rate, the very same 'forms' which constitute the essen-

---

1 As I have pointed out, the epistemological need for justification is a more recent phenomenon arising under the pressure of 'Neo-Platonist' and mechanistic ontologies (cf. ch. I).

tial ontology of the world—the objects and their essential attributes or qualities—also *in-form*[1] the percipient in a literal non-metaphorical sense. In seeing a red object the observer's eye and mind become red in a way. Of course, there were other schools of thought, those that held that the real objects of perception are not physical qualities but mathematical and spatial properties. The theories of vision put forward by these *perspectivists* differed accordingly. But at this stage it is important to note that they, too, shared what I will call the *identity theory of perception* according to which the essential characteristics of the world are identically present in the knower and the known. Thus, psychological and intellectual processing of the given was not needed to *transcend* sense perception but to render it *intelligible*. For whatever was transmitted to the senses in whatever complicated and as yet ill-understood way, it contained the real essence of the objects of perception and not mere images thereof. To disregard the evidence of the senses, therefore, or to distrust it systematically, was to distrust the essential manifestations of nature which she, nature herself, by some physical process delivered at our very doorsteps.

The implicit premises of medieval theories of vision, therefore are:
(1) The Identity Theory of Perception: what is conveyed to the senses is reality itself and not images or, worse still, mere signs thereof. The veil of appearances which would later separate man from his world was not yet drawn.
(2) The peripheral 'response' as such conveys solid sensory *knowledge* and not merely enigmatic *information* in the modern sense of cues or data about a necessarily elusive external world. This knowledge is fallible, to be sure, but only under nonstandard conditions which are limited in number and largely specifiable in advance. Consequently, most illusions of the familiar kind are 'naturally to be expected'. Perceptual error, therefore, is a local phenomenon, not a global disaster.

Clearly, in such a context Cartesian doubt would be a freak case of philosophical pathology. Medieval epistemology was not yet plagued by that obsession with cognitive absolutism characteristic of later philosophical theories. Because it regarded vision as the reception and absorption of forms either by direct contact (mathematical theory of extramission) or by literal in-form-ation (Peripatetic and Baconian-Alhazenian theory of intromis-

---

1 The origin of this now modern term clearly betrays Aristotelian lineage. For a transitional stage in the development of the term and its associated concept, cf. Avicenna, *Liber canonis de medicinis cordialibus*, III, III, I, 50 (Venice, 1555). Here the translator's term '*informatio*' is already used in the sense of the impression of (formally identical) *images*.

sion) the problem of cognitive achievement (or the overall correspondence between distal and central variables) could be seen as solved in principle once the extra-organismic leg of the perceptual process had been clarified. For then the explanatory postulate implicit in Scholastic epistemology would have been satisfied: knowledge—or the possession of true ideas about the physical world—presupposes some kind of state or property which is identically present in the knower and the known.

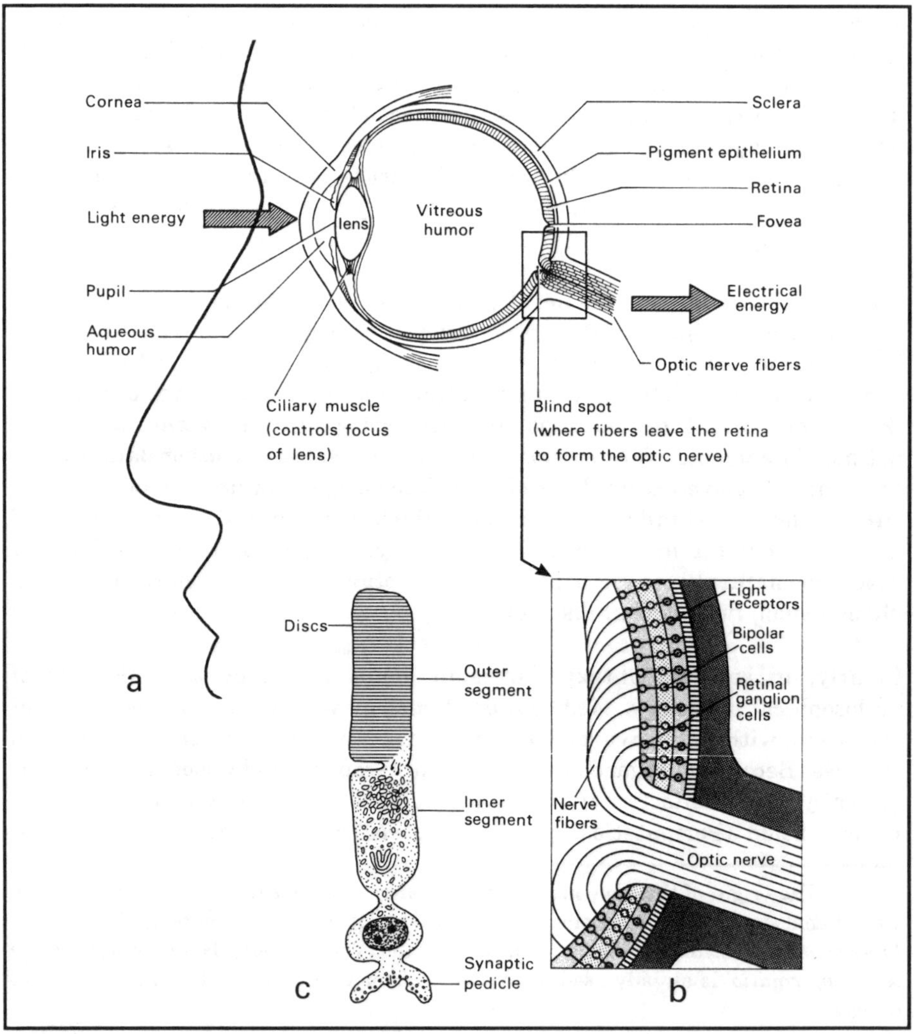

***Fig. 2 — The eye as transducer.** Contemporary view of the anatomy of the eye, here portrayed as a contrast to earlier views of visual perception. The receptors, located in the retina, a complex network of neurons lining the back of the eye, transform optical energy into electrical impulses which are propagated as action potentials along the optic nerve. In (b) the retina's 'vertical' organization is shown. Note that the light-sensitive receptors of the eye are tucked away* inside *the retina. Thus light has to pass through several (albeit nearly-transparent) layers of cells and blood vessels before it can stimulate the receptors. A visual receptor is shown in detail in (c). Electrico-chemical transduction is initiated in the tiny* lamellae *(or discs) of the outer segment, which contains a light-sensitive molecule called* visual pigment, *consisting of a protein called* opsin *and a light-sensitive substance called* retinal. *The retina is capable of highly sophisticated visual analysis all by itself. In view of its computational resources as well as its anatomical origin (the retina develops in the foetus as an outgrowth of embryonic brain tissue) it is not inappropriate to regard the retina itself as a kind of visual brain.*

It is one of the greatest ironies of the history of ideas that the (Keplerian) near-solution of *the* problem of vision as construed by the Aristotelian research program at the same time necessitated the radical abandonment of the latter's explanatory scheme: the Scholastic identity theory had to be dropped in favor of a scheme of mechanical explanation. Apparently, our perceptions—and ultimately our intellectual ideas—are not the result of some process of contagion by 'sensible species'. Rather they are somehow caused by mechanical motions of physical light propagated along lines capable of mathematical analysis and specification. The theory of coherent forms entering our eyes as physical wholes is replaced by a theory of punctiform analysis whereby objects are 'reconstructed' at the sensitive surface of the retina from numerous 'bits of mechanical information'.[1] The traditional identity theory of perception no longer works. For nothing can be more dissimilar than mechanical motions and pressures on the one hand and the physical qualities of Aristotelian science as revealed by the senses on the other. *Thus, paradoxically, perception—and, consequently, knowledge in general—became profoundly problematic once the traditional problem of perception had been solved.*

---

1 This theory was anticipated, but not radically worked out along mechanicist lines, by Alhazen. In the Baconian synthesis of the Scholastic era, however, it was even further removed from mechanical elaboration. (See ch. IV, sec. 3-4).

Knowledge of sensory function now appeared altogether inadequate to explain the overall correspondence between our ideas on the one hand and the physical qualities defining the objects on the other. In fact, that very correspondence itself became a legitimate matter of controversy. For the old question of epistemology: where do our ideas, or concepts, come from, or what explains their objective validity, if any—a question which was marvelously explained within the framework of Aristotelian empiricism—now re-emerged more vexing than ever before. As our ideas apparently do not come into the mind from outside in sense perception (since sensations and physical qualities are altogether unlike each other) they may at best be regarded as some kind of mental *representations* of the surface characteristics of the outside world. Somehow the mind must *represent* to itself the qualities of external objects. But the big question now is: how is this process—the transition from external qualities to internal representations—objectively controlled?

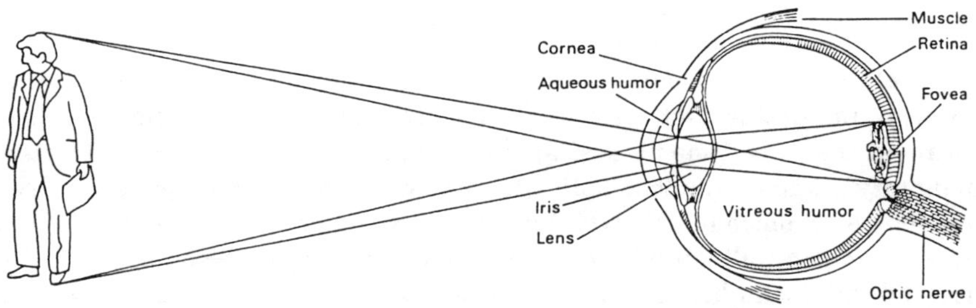

***Fig. 3 — Modern conception of the dioptrics of the eye.** Shown here as a contrast to earlier views. The inversion of the optical image within the eye constituted a major obstacle to pre-Keplerian theory formation in physiological optics, with most investigators postulating special processes either preventing inversion or else correcting it so as to preserve an erect image being transmitted to the* sensus communis.

Thus the breakdown of Aristotelian empiricism and its replacement by the mechanistic hypothesis led at once and quite naturally to the philosophical demand for *justification* of our sensory knowledge. Implicit is the radical conception of knowledge as something of our own making for which we ourselves may be responsible. Furthermore, the epistemological dualism between ***knowledge*** and 'mere' (psychophysiological) ***information*** was inseparably linked to the ontological dualism of body and mind. A third subject of modern philosophical inquiry since Descartes concerned the psychological (and intra-organismic) question of the adequacy of the mind as an information-processing device. Hence the increasing importance of learning theories and of information-theoretic accounts of (conscious or unconscious) cue utilization.

For the time being I will only make a few brief comments pertaining to this transition from the Aristotelian identity theory to the mechanistic hypothesis. In the first place it demonstrates that the splitting-up of empirical domains may be a necessary condition of important scientific progress. This phenomenon deserves more attention than it has been given so far. Indeed, modern philosophy of science (from Whewell's theory of consilience and Carnap's physicalism down to contemporary reconstructions of theoretical growth) tend to represent domain *fusion* as the only true sign of scientific advance. Newtonian mechanics is usually held up as the inspiring classical example for this as it unified the traditional domains of the celestial and the terrestrial, the superlunary and the sublunary spheres by subsuming these separate realms under one universal explanation. But in the heart of that very same mechanistic program we find a budding theory of perception whose progress was conditioned precisely by its *separating* optical from epistemological, and physiological from psychological questions which formerly had been regarded as belonging to one and the same scientific domain.

Secondly, the progress of the theory of perception is better measured in terms of the numerous novel and 'deep' questions it made possible than in terms of the partial (physiological) solution of the problem of perception which raised those questions. The 'small' step towards the final solution of the dioptrics of the eye which after its long preparation during the Middle Ages remained to be taken by the end of the 16th century—and which Kepler took in fact—turned out to involve a conceptual revolution which paved the way for a scientific research program of immense scope. In Descartes, in his *Dioptrique*, in his theory of bio-automatism and in his formulation of the mind-body problem we see this wide-ranging program already in full operation while far-reaching solutions and methodological suggestions are philosophically argued for which will dominate the history of science for decades and modify it forever.

In the third place this conceptual revolution clearly shows that philosophical and empirical questions are directly and almost inextricably intertwined. For on the one hand, the *epistemological* postulate of the identity theory provided the matrix for the direction of research, for the formulation and eventual resolution of all kinds of empirical (optical) problems. On the other hand, the new *mechanical* model of the eye as an optical instrument necessitated radical changes in the theory of knowledge. The later issue of the impact of mechanical models on the development of epistemological (psychological and perceptual) theories during the 17th and 18th centuries will be discussed at a later stage of the argument.[1] First I want to provide evidence for the former thesis that a 'typically' epistemological conception served as a 'negative heuristic' for the theory formation in various early optical traditions. If our (global) reconstruction of the early history of optical theory in terms of a Lakatosian research program sheds light upon otherwise unintelligible details of concrete developments in the theory of vision during late antiquity, and especially during the early (Arab) Middle Ages and the Scholastic era, then this may also serve as a partial corroboration of the historical adequacy of the methodology of scientific research programs.

---

1 Cf. ch. VI.

III

THE FORMATION OF COMPETING OPTICAL TRADITIONS IN EARLY AND LATE ANTIQUITY.

1. Rival research programs in classical theories of visual perception.

No doubt the greatest medieval scholar in the field of optics was the Arab Ibn al-Haytham (fl. ca. 1000 AD), known in the West, and immensely influential, under the name of Alhazen or Alhacen. Before his time three clearly distinguishable optical traditions existed in the Arab academic world whose theoretical foundations and ideas concerning even such fundamental questions as the very aim of optics were so vastly different that sometimes one gets the impression that they represented altogether different areas of research rather than competing research programs for one and the same domain of phenomena.

Moreover, although these three traditions displayed a certain amount of methodological overlap in peripheral areas of common interest, nevertheless with regard to their respective focal areas they can largely be defined in terms of different academic disciplines with different areas of application. Thus optical thought in the Galenic-Stoic tradition prevailed primarily in medical circles (Hunain ibn Ishaq, 'Ali ibn 'Isa, 9th century A.D.). Its research especially focused upon physical, physiological and anatomical aspects of vision. By contrast, the optical tradition stemming from Euclid, Hero and Ptolemy flourished principally among mathematicians (such as al-Kindi, 9th century A.D.). Its theory of vision concentrated in particular upon a geometrical treatment of perspective and of the spatial aspects of vision in general. In the third place, Aristotle's theory of vision became increasingly popular among (natural) philosophers such as Avicenna (Ibn Sina, 980—1037) and Averroes (Ibn Rushd, ca. 1126—ca. 1198). This theory particularly emphasized the 'natural' causal factors, both physical and psychological, which it declared operative in the perceptual process.

Thus one might think that these optical traditions, rather than constituting competing schools of thought, in fact represented complementary areas of scientific interest. However, such an interpretation would be profoundly anachronistic. This is clear not only from its failure to provide an adequate internal[1] explanation of the ongoing discussions which until the time of Alhazen continued to flare up between the representatives of the various traditions, but also from the fact that Alhazen's eventual 'solution' did not consist in splitting up the several traditions into separate disciplines with corresponding domains but precisely in integrating in a unique way selected elements from those seemingly incompatible optical traditions within the framework of a novel and comprehensive theory of vision. What had made these former theories into apparently incompatible and actually competing theories of vision was precisely their failure to make the modern distinction between geometrical, physical, physiological and psychological aspects of the visual process. All these 'optical' theories primarily aimed at explaining *visual perception* as an integrated process. The eye and the faculty of vision, optical stimulation and visual perception, these were all regarded as essentially one in the act of vision. Consequently, it was quite evident that the eye could only be studied as the seat of the visual faculty and not as a dead optical instrument merely consisting of several refracting media. This ancient theoretical conception is suggested by 'common sense' and, as it were, frozen into linguistic practice in the meaning of such words as the Greek ὄψις (opsis), the Latin *visus*, or the Arabic *baṣar*.[2]

In fact, the methodological requirement of minimal adequacy went much further than this: any optical theory should at least be able to explain how the essence of the visible world is wholly retained in visual perception, or how the primitive (phenomenal) facts of visual perception are identically present in both the percipient and the perceived. Even in formal respects, therefore, 'optical' theory during the Middle Ages differed radically from modern optics inasmuch as it regarded optical and perceptual questions as systematically connected. This implied that the 'psychology' of perception could dictate an anatomical arrangement.[3] On the other hand, one can find

1 Cf. I. Lakatos, 'History of Science and its Rational Reconstruction', in *P.S.A. 1970, Studies in the Philosophy of Science* 8, R.C. Buck and R.S. Cohen (eds.), (Dordrecht, 1971), pp. 91-135.

2 As A.I. Sabra points out, Alhazen's *baṣar*, like the Greek ὄψις (*opsis*) and the Latin *visus*, means both eye and sight (or sense or faculty of sight). Cf. A.I. Sabra, 'Sensation and Inference in Alhazen's Theory of Visual Perception', in *Studies in Perception*, P.K. Machamer and R.G. Turnbull (eds.), (Columbus, 1978), p. 180.

3 Incidentally, in the case of Muslim scholars in the field of optics it is in principle possible to offer an external explanation for this, from a modern point of view, curious

epistemological notions being defined in optical terms.[1] Introspective knowledge of visual experience on the one hand, and on the other the epistemological postulate of the formal unity of the percipient and the object in the act of perception constituted the core data, or the disciplinary matrix, to which optical theory formation had to accommodate itself. Consequently, 'optics' was a highly 'naturalistic' discipline: unlike its 17th century successor its pattern of reasoning did not yet presuppose a deep dichotomy between two epistemologically utterly distinct realms.

As was mentioned above, of the optical theories of antiquity only the Aristotelian, the Stoic-Galenic and the Euclidean-Ptolemaic theories were developed into full-fledged comprehensive research programs. In the following I will briefly discuss the classical ancestry of these optical traditions.

2. The Aristotelian theory of vision.

With the Atomists and with Plato, Aristotle recognized the need of some kind of physical intermediary between the percipient and the visible object. For if visual perception involved no more than a purely abstract relationship (such as between equals) then

> there would be no need [as there is] that either [the beholder or the thing beheld] should occupy some particular place; since to the equalization of things their being near to, or far from, one another makes no difference.[2]

On the other hand the idea of a material effluence either from the visible object to the eye or vice versa is absurd for various reasons.[3]

---

state of affairs in view of the fact that the Islam prohibited dissection. Thus (one might argue) for their anatomical theories the Arabs depended either upon the Galenic heritage or else upon (psychologically inspired) speculation.

Yet this explanation has weaknesses of its own. For actual anatomical research could equally well be carried out by non-Muslim contemporaries. It is no accident that, for example, the Arab ophthalmologist Hunain ibn Ishaq who worked extensively in the field of ocular anatomy and physiology, was a Nestorian Christian.

1 Cf. below, ch. IV, 3, p. 53.

2 *De sensu* 6, 446b 12 ff., *The Student's Oxford Aristotle*, W.D. Ross (ed.), (New York, 1942), Vol. III. All subsequent references to works by Aristotle are taken from this edition unless otherwise indicated.

3 Cf. ibid. 2, 437a 18-438b 2. Also cf. *De anima* 2, 7, 419a 12-21.

Unlike his predecessors, therefore, Aristotle stresses the need of an optical medium between observer and visible object. What then, is this optical medium? It is the diaphanous or the translucent, i.e. a 'nature' or power common to all things to a greater or lesser degree but residing especially in air, water and many solid objects commonly called transparent. Indeed, the transparent is not visible itself, but it is that through which we see. The transparent consists in the power of a substance S (characterized by transparency) to communicate the colors of bodies on the other side of S to a recipient on this side of S.[1]

The transparent is excited to actuality by the influence of fire or something resembling the 'uppermost body'. Light (φῶς) is the actualized state of the (potentially) transparent. Thus, light is variously called the activity of the transparent, or, again, the presence of fire or of something resembling fire in what is transparent. But light is not itself fire nor any other kind of body whatsoever. More specifically, light cannot be any kind of material emanation either (as with the Atomists). If it were, it would again itself be a kind of body. However, this is impossible, for two bodies cannot be present in the same place.[2] Thus, light is an (actual) state of the transparent in virtue of which objects separated from the observer by the (potentially) transparent medium become visible. Consequently, the 'propagation' of light is instantaneous. Unlike the transmission of corporeal particles or of energy it takes no time.[3]

Aristotle defines color (χρῶμα) either as the limit of the transparent in a determinately bounded body[4] or else as the surface layer of visible objects with the power of producing qualitative changes in the actually transparent medium.[5] This is the core idea. Thus the co-activity of light and color moves the transparent medium and this, extending continuously into the interior of the eye, acts upon the sense organ and produces there the same specific qualitative change.[6] However, the watery substance of the eye *is* at the same time the proper seat of the faculty of vision which is wholly receptive to light and color. *Thus the eye assumes the qualities of the visible object and*

1 *De sensu* 3, 439a 21-25; *De anima* 2, 7, 418a 29-418b 9.
2 *De anima* 2, 7, 418b 9-19.
3 *De sensu* 6, 446b 22-447a 17.
4 Ibid., 3, 439b 11f.
5 Ibid., 3, 439a 27-34.
6 *De anima* 2, 7, 419a 8-21.

literally becomes what it sees. The actuality (ἐνέργεια) of the sensation and its object is the same, only their being (εἶναι) or location is different.[1]

Clearly, in Aristotle the identity theory is most explicitly articulated. An interesting and far-reaching consequence of this theory, according to Aristotle, is the denial of the existence of invisible magnitudes and imperceptible time intervals. Obviously, what is at stake here is the theoretical relevance of mathematics for the study of physics. Indeed, Plato denied this relevance as well. This runs counter to popular contemporary ideas about an alleged connection between Platonism and mathematical physics in the historical origins of modern science.[2] By contrast, the existence of imperceptible intervals of time and space (or matter) became a principle of vital importance for Descartes who had literally identified mathematics and physics. In direct opposition to the atomists he maintained that there is no finite limit to the divisibility of matter.[3] For, according to Descartes, the indefinite divisibility of matter is an absolute condition of the total and radical geometrization of nature.

3. The Stoic-Galenic Tradition.

If the intelligibility of the Aristotelian conception of the eye as the seat of visual perception already requires a rather drastic *Gestalt* switch of modern

1 Cf. *De anima*, 1, 3, 407a 7; 2, 5, 418a 3-6; 2, 12, 424a 16-24; 3, 1, 425b 17; 3, 4, 429a 28; 3, 8, 432a 1.

2 The radical notion of the fundamental applicability of mathematics to the physical world is a novel idea gradually prepared in the development of medieval optics (as I will argue in ch. V; also cf. ch. III, sec. 4) to be consciously embraced only by the founders of modern science. Its radical content as well as its complicated history militate against its being simply classified among the ancient wisdoms of Plato. At any rate, in a strict sense this notion is not Platonic in origin at all. It is no surprise, therefore, to find intellectual historians recently making a compelling case against the myth of popular historiography according to which Galileo was a Platonist. Galileo certainly did not share Plato's view of the relationship between the intelligible and the sensible. Cf. Dudley Shapere, *Galileo: A Philosophical Study*, (Chicago, 1974) and Hans Blumenberg, 'Pseudoplatonismen in der Naturwissenschaft der frühen Neuzeit', *Akademie der Wissenschaften und der Literatur, Abh. der Geistes- und Sozialwissenschaftlichen Klasse*, (Mainz, 1971), nr. 1, pp. 3-34.

3 Cf. R. Descartes, *Principles of Philosophy*, II, 20 (HR I 264); II, 34 f. (HR I 267).

man's common sense, the Stoic-Galenic notion that the sensory power is somehow communicated to the surrounding air between observer and object puts an even higher demand upon our weak inert imagination. However, the Galenic theory is by no means absurd. At any rate, it is no more absurd than any theory that attributes sensitivity to matter.

In Galen's view the problem of vision allows no more than two alternative theoretical solutions: either the visible object communicates its specific character to us by transmitting something from itself towards us; or else the communication is achieved because our sensory power extends towards the perceived object.[1] Since the former alternative has several untenable consequences only the latter solution remains. Its elaboration involves the Stoic conception of *pneuma* (πνεῦμα), an all-pervasive agent composed of a mixture of air and fire.[2] The optical pneuma flows from the seat of consciousness, the hegemonikon (ἡγεμονικόν), through the (hollow!) *nervus opticus* (or optical nerve)[3] to the eye. Upon its emergence from the eye it immediately combines with the adjacent air and assimilates it instantaneously (just as sunlight only has to touch the upper limit of the air in order to transmit its power directly to the whole).[4] This double instantaneous assimilation of the air to the sensitive pneuma as well as to the light from the sun transforms the air itself into a homogeneous instrument of perception for the eye (just as the nerve is for the brain[5]) which under sufficient illumination is capable of discriminating its proper sense-objects. Thus, with the air itself percipient the sensory power literally extends all the way to the visible object.

---

1 Claudius Galenus, *De placitis Hippocratis et Platonis*, Libri novem, Vol. I, Prolegomena critica, textum graecum, adnotationem criticum versionemque latinam continens, Iwan Mueller (ed.), (Leipzig, 1874), 7, 615, 9-13 (K 5, 618). Subsequent references will be to this edition of Galen's *De placitis*, to be followed, in parentheses, by the corresponding reference in the Kühn edition of Galen's works.

2 Cf. S. Sambursky, *Physics of the Stoics*, (London, 1959), pp. 1-11. On the composition of *pneuma*, cf. e.g. Alex. Aphr., *De anima*, 26, 16; *De Mixt.*, 224, 15.

3 On the discovery of the nerves and the persistent theory of the hollow nerves, cf. Friedrich Solmsen, 'Greek Philosophy and the Discovery of the Nerves', *Museum Helveticum* 18 (1961) 150-167, 169-97; Edwin Clarke, 'The Doctrine of the Hollow Nerve in the Seventeenth and Eighteenth Centuries', in *Medicine, Science, and Culture: Historical Essays in Honor of Owsei Temkin*, Lloyd G. Stevenson and Robert P. Multhauf (eds.), (Baltimore, 1968), pp. 123-41.

4 *De placitis* 7, 615, 3-7 (K 5, 617); 7, 616,14—617,4 (K 5, 619).

5 Ibid., 7, 623, 2-7 (K 5, 625).

4. The geometrical tradition.

The origin of the third, geometrical, tradition of medieval optics goes back to the work of Euclid, Hero and especially Ptolemy. As a result of its Platonic inspiration (Euclid was a pronounced Platonist[1]) this research tradition is initially marked by its close association with pure mathematics. Of course, in Plato's ideal true theoretical science is deemed incomparably superior to mere empirical knowledge (on the level of culinary art) or even to practical science based upon technical insight and ingenuity. It is not that these latter activities can only produce apparent 'knowledge', i.e. knowledge based upon no more than the subjective appearance of things. On the contrary, the intrinsic pseudo-character of their cognitive products has a much more radical, viz. *ontological*, ground. For, surely, these activities are concerned with *objective* material phenomena. Yet, these phenomena constitute at best severely impoverished, indeterminate and at any rate incalculable replicas of their eternal immutable archetypes in the realm of ideas. True knowledge (ἐπιστήμη), which is as universal and immutable as its object, is possible only with respect to this latter reality. The sensory world, however, is not a perfect systematic projection of the sublime ideas but an inaccurate realization of it into recalcitrant matter. Thus, the archetypal ideas cannot simply be recovered by means of some formal operation with respect to sensory knowledge. Consequently, knowledge derived from the study of material phenomena is doomed to remain inadequate, imperfect and fragmentary, perhaps capable of stumbling upon accidental rules of thumb, but not of formulating universal laws whose validity is necessary. The unenlightened mind who regards this knowledge as 'true' knowledge resembles the poor art collector who holds his kitsch copy for a true Vermeer.

This Platonic philosophy of science obviously stimulated a thorough mathematization of science. It is no accident that during antiquity those disciplines developed most rapidly whose empirical domains lent themselves most easily to mathematical characterization and in which, furthermore, a minimum of observation and experiment already sufficed for comprehensive theory formation.[2] But it is important to realize that the underlying philosophy of science implied a methodology and ontology with respect to these 'classical' sciences totally unlike what might be suggested by

---

1 Cf. E.J. Dijksterhuis, *De mechanisering van het wereldbeeld*, (Amsterdam, 1950), p. 56.

2 Cf. T.S. Kuhn, *The Essential Tension*, (Chicago, 1977), pp. 37-8.

the misleading name of "classical *physical* sciences".[1] For in the first place these sciences are not primarily concerned with material phenomena, let alone with the idea that these might be susceptible of exact mathematical description. Rather, what is being described or investigated is an 'ideal' nature, or nature as it would have been if matter had not distorted its perfect conception. Thus, all classical 'sciences' are actually branches of pure mathematics. Just as 'geometry' (so-called) had ceased to be terrestrial geometry, so Euclid's *Optica* did not primarily claim to be a detailed theory of natural vision. Similar remarks apply to the other classical disciplines. Being purely mathematical in orientation their principal objective was not to formulate formal models of meticulously studied and possibly experimentally explored natural domains which were expected to be their realizations. Rather, the natural world only served as the occasion for contemplating the intuitions, slumbering within our minds, of the purely conceptual archetypal ideas. It is of these latter ideas alone that the various mathematical theories are the respective formal expressions and not, or not primarily, of their crude and clumsy material counterparts. Knowledge of these latter entities represents no more than a practical knowledge of nature which cannot even serve to refute the theoretical mathematical sciences.[2] Ideally, and originally, a clash between the mathematical and the empirical 'sciences' was thus inconceivable as they described and explored different scientific domains while employing methods suited to the different ontological character of their respective fields of inquiry.

---

1 Ibid. p. 35 (emphasis added). Kuhn especially classifies astronomy, harmonics, mathematics, optics and statistics as belonging to this category.

2 Even Ptolemy whose merit it was to have realized that optics, notwithstanding its profoundly geometrical character, yet requires a method essentially different from that of pure geometry [cf. A. Lejeune, *Euclide et Ptolémée. Deux stades de l'optique géométrique grecque*, Université de Louvain, Recueil de Travaux d'Histoire et de Philologie, ser. 3, fasc. 31 (Louvain, 1948), p. 41], nevertheless chose to ignore, or to smooth over, whatever discrepancies he found between his careful and often ingenious measurements on the one hand and the exigencies of his a priori theoretical models on the other, always favoring the latter (e.g. in his determination of the angular dimension of the visual cone (cf. *ibid.* pp. 41-51) as well as in his drawing up of the tables of refraction [cf. id., 'Les tables de réfraction de Ptolémée', in *Annales de la Société Scientifique de Bruxelles*, ser. 1, vol. 60 (1946), pp. 95 ff.]. Having said this much, however, I should add that there are also unmistakable methodological tensions operative in Ptolemy's work as he is caught between his loyalty to the archetypal Euclidean conception of science and his own experimental inclinations. These ambiguities are especially manifest in his work on diplopy and on the horopter [cf. id., 'Les recherches de Ptolémée sur la vision binoculaire', *Janus* 47 (1958), pp. 79-86)]

However, in the course of time a gradual 'physicalization' of the classical sciences takes place. During the Hellenistic period it is already clearly demanded that purely mathematical models should also be subject to purely physical ('terrestrial') interpretations. On the one hand existing mathematical theories are increasingly enriched with physical details. 'True' science more and more aims at incorporating observed physical phenomena into fundamental mathematical theories whose basic principles are unshakable. In other words, instead of designing and developing axiomatized theories with respect to ideal or idealized domains, classical science begins to aspire to offer mathematical explanations of physical phenomena and, if necessary, to accommodate the mathematical details to the observed facts. In certain fields of inquiry, e.g., in mathematical astronomy, this tendency leads to what is now interpreted as instrumentalism, or as 'saving the appearances'. On the other hand, physics begins to assimilate parts of what was originally presented as pure mathematics by offering physical explanations of mathematical postulates. Thus, according to Damianus[1], Ptolemy showed by experimental means what Euclid had simply assumed as an axiom, viz., the rectilinear propagation of visual rays. Again, al-Kindi felt a similar need of physical demonstration. He attempted to show the rectilinear propagation of luminous rays, which he, like the Greek opticians before him[2], held to be identical with visual rays (or at least as far as their mode of propagation was concerned).[3]

However, this increasingly *ontological* role of mathematics amounted in none of the classical sciences to a full-fledged mathematical physics, *except in optics*. For unlike in astronomy, where the existence of a mathematical research tradition alongside a physical one could be interpreted as the result of a corresponding diversity of fields of inquiry and of target tasks (mathematical astronomy as celestial kinematics and physical astronomy as natural philosophy and cosmology), the physicalization of geometrical optics necessarily gave rise to a *competing* mathematical research program for the domain of visual perception. Thus, whereas the conflict between mathematics and physics, of such momentous significance for the birth of modern science, came to a head in *astronomy* not earlier than the sixteenth century when a mathematical astronomy with physical pretensions superseded a natural philosophical cosmology; and whereas in *mechanics* the conflict led

---

1 Damianos, *Schrift über Optik*, R. Schoene (ed.), (Berlin, 1897), p. 4, 7-20.

2 Cf. A. Lejeune (1948), p. 62.

3 *De aspectibus*, propositions 1-3, in 'Alkindi, Tideus und Pseudo-Euklid. Drei optische Werke', A.A. Björnbo and S. Vogel (eds.), *Abhandlungen zur Geschichte der Mathematischen Wissenschaften*, vol. 26, 3 (Leipzig/Berlin, 1912) pp. 4 ff.

to important new developments not earlier than the seventeenth century when a mathematical mechanics and dynamics finally drove the physical doctrine of natural and unnatural motions out of its traditional position of dominance; that very same fundamental conflict occurred in *optics* as early as the early Middle Ages in Islamic visual theory where in the Alhazenian synthesis it resulted in a predominantly mathematical theory of vision. In chapter V I will discuss this latter point more fully. First the ground has now been prepared for a brief exposition of Euclid's and Ptolemy's classical optical theories.

In Euclid the methodological objective to transform optical problems into purely geometrical ones is still the most pronounced. He axiomatizes the domain of optics by means of seven geometrical postulates[1] and apparently wants to steer clear of any ontological interpretations of a physical, physiological or psychological kind. That is to say, those aspects of visual perception not immediately reducible to pure geometry are not of genuine 'scientific' (i.e., mathematical) interest.[2] Thus it appears as though Euclid's *Optica* is far removed from the identity theory of perception which during the Middle Ages would guide all optical research. However, this is only apparently so. For in the first place the *Optica* does not provide a geometrical optics in the modern sense of the word but rather a *geometrical theory of visual perception*. And, secondly, as a corollary to the first point, Euclid's optical postulates clearly involve a scientific ontology with respect to the entire physical and psycho-physical chain of the visual process.

The theory of visual perception—for in essence that is what it is—which can be extracted from Euclid's postulates runs as follows. The eye touches the environment by means of rectilinear visual rays which diverge indefinitely from the eye and jointly constitute a cone, the apex of which is in the eye while its base is located upon the surface of the perceived objects. The apparent magnitude of the perceived object is a direct function of the angle under which the object is perceived. The apparent altitude is determined by higher (or lower) visual rays. Analogous geometrical functions apply to the apparent orientation. Finally, the visual definition of the object is determined by the number of the angles under which it is perceived.

---

1 Euclid, *Optics*, Definitions 1-7, in *A Source Book in Greek Science*, M.R. Cohen and I.E. Drabkin (eds.), (Cambridge, 1966), pp. 257-8.

2 For example, Euclid pays no attention whatever to the essential roles of light and color in vision. All he needs for the construction of a deductive theory of the visual perception of space is the assumption of visual rays. Apparently the spatial information these rays convey is assumed to be affected in no way by these two factors which otherwise so manifestly determine visual sensation.

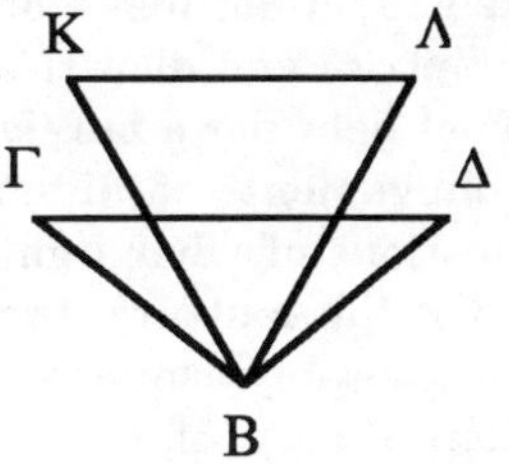

***Fig. 4 — Euclid's principle** that the apparent size of a visible object is proportional to the angular separation between the visual rays grazing its extremes, i.e. the angles $\Gamma B\Delta$ and $KB\Lambda$ subtended at the eye at B.*

After Euclid the process of physicalization within the tradition of research he founded advanced even further, no doubt precipitated in part under the influence of Aristotelian empiricism. Consequently, the theoretical relations with other research programs in the field of vision assumed a more competitive character. As I indicated above, in astronomy an eventual clash could be avoided by a timely definition of complementary areas of research. In optics Hero of Alexandria (1st century A.D.) attempted to achieve a similar intellectual distribution. He divides the theory of vision into optics (i.e., the science of vision proper), dioptrics (i.e., the science of refraction) and catoptrics (i.e., the science of reflection)[1] and regards the former area as adequately dealt with by Aristotle while claiming only the latter areas for himself.[2] Apparently he sees his studies—or attempts to represent them—as

1 Hero, *Catoptrics*, 1, in Cohen and Drabkin (1966), pp. 261-2. In his *Commentary on the First Book of Euclid's Elements* the Neoplatonist Proclus (A.D. 410-485) divides optics into optics proper (including the theory of perceptual illusions), general catoptrics (which is concerned with the various ways in which light is reflected) and scenography (or applied perspective); cf. Proclus, *A Commentary on the First Book of Euclid's Elements*, I, 40, Glenn R. Morrow (tr.), (Princeton, 1970), p. 33.

2 In Aristotle similar suggestions of a complementarity thesis for possible subdivisions of optics can be found occasionally. He himself nowhere discusses perspective, neither in his general nor in his detailed accounts of perception, nor does he anywhere make use of a (Euclidean) visual cone or pyramid. Yet as the long-term colleague of Eudoxus in Plato's Academy he may have thought of (branches of) optics along lines similar to those later adopted by Euclid and especially by Hero. Thus in a context where he attempts to distinguish physics from mathematics he writes:

> "Similar evidence is supplied by the more physical of the branches of mathematics, such as optics, harmonics, and astronomy. These are in a way the converse of geometry. While geometry investigates physical

complementary to those of Aristotle. Yet, how idle is this evasive attempt! Hero's seemingly modern allocation of intellectual tasks should not make us blind for its utterly unmodern character. For Hero's catoptrics and dioptrics are neither purely geometrical models of the propagation of light nor a purely fictitious kinematics of the ideal form of the movement of light *independently of* such physical and psychophysical questions of what light really is and whence it comes and how it is perceived. On the contrary, the geometrical theory of mirrors of Hero's *Catoptrica* is inextricably connected with the visual ray ontology of Euclid's extramission theory of visual perception *and the latter is incompatible with Aristotle's theories of light and vision*. Hero's research problems are entirely dictated by this theoretical perspective. Thus he wonders (with others) why *the visual rays proceeding from his own eyes* are reflected by mirrors and why the reflections are at equal angles.[1] Or he argues that the *visual rays* move with infinite velocity because when we *open our eyes* and *look up at the sky* no interval of time is required for the *visual rays* to reach the sky.[2] Furthermore he formulates an elasticity theory of reflection.[3] Thus it is evident that for him the visual rays are not mathematical fictions but material entities.

Also working within the Euclidean tradition Claudius Ptolemy, too, regarded the visual ray as something physically real. According to a later commentator, Simeon Seth, Ptolemy says in his *Optica* that "the visual pneuma is something of ether, belonging to the quintessence".[4] Note first of all the specifically Aristotelian accretions to the Euclidean tradition as the latter progressively undergoes the process of physicalization mentioned above. Secondly, it is obvious that for Ptolemy, too, the notion of visual radiation is not a purely mathematical, and thus purely abstract, imaginary or at any rate non-physical, concept. For the association of visual radiation with 'external' light is here clearly articulated in view of Aristotle's association of the quintessence, or the ether, with light.[5]

Unlike Euclid Ptolemy did not represent the visual energy proceeding from the eye as composed of discrete diverging rays whose mutual distance in-

---

lines, but not *qua* physical, optics investigates mathematical lines, but *qua* physical, not *qua* mathematical." (*Physics*, II 194a 7-11).

1 *Catoptrics* 2, in Cohen and Drabkin (1966), p. 263.

2 Ibid.

3 Ibid., pp. 263-4.

4 Quoted in A. Lejeune (1948), p. 65.

5 Cf. ibid. For Aristotle's 'ethereal' theory of light, cf. *De anima*, 2, 7, 418b 14-19.

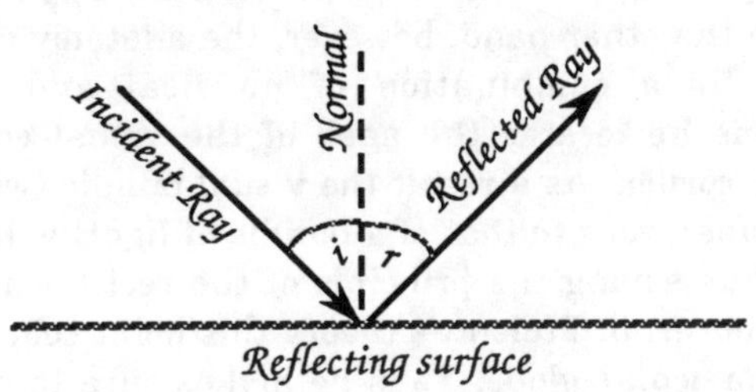

Fig. 5 — Reflection of light rays. *The angle of incidence is equal to the angle of reflection. Angles are measured from the normal, i.e. the line perpendicular to the plane of the reflecting surface at the point of incidence.*

creased with their distance from the eye but rather as constituting a continuous visual cone.[1] No doubt this aspect of his geometrical model was

---

1 Obviously, this mathematical representation, though more in accord with the facts of phenomenal experience, is hardly compatible with the realist tendency to interpret the theoretical concept of the visual ray as something physically real. The notion that an infinity of distinct rays is contained in each and every visual angle cannot be reconciled with the implied 'ethereal' nature of these rays, let alone with ascribing to them definite mechanical properties. The tension building up under the influence of the progressive physicalization of 'perspectivist' theories of optics in late antiquity and during the early (Arab) Middle Ages thus articulates at a relatively early time the profound epistemological problem of applied mathematics while preparing the way for a full-fledged philosophy of mathematical physics. For the only way Ptolemy can resolve the above paradox, it seems, is the one suggested by Lejeune:

> "...that model [i.e., the mathematical model involving an infinity of discrete visual rays so as to assure the apparent continuity of the discrete visual field] must remain geometrical. To materialize, or, if you wish, to arithmetize it, is to falsify it."
> ["...cette représentation doit rester géométrique. La matérialiser, ou, si l'on veut, l'arithmétiser, c'est la fausser."] (A. Lejeune (1948) p. 83).

Alkindi adopts a similar solution which he, in his reverence for the old master, ascribes to Euclid himself. Thus, he argues, although the visual radiation from the eye is really continuous, its overall shape can be mathematically represented as an infinity of discrete geometrical lines, because

> "the boundaries [i.e., the lateral surface] of the cone-shaped figure impressed in the air by the visual power proceed with the rectitude of straight lines separated by intervals." [*De aspectibus*, A.A. Björnbo and S. Vogl (eds.) p. 15; cf. David C. Lindberg, 'Alkindi's Critique of Euclid's Theory of Vision', *Isis*, 62 (1971), esp. pp. 478- 81].

suggested by psychological considerations (viz. the apparent continuity of the visual field). On the other hand, however, the anatomy of the eye in Ptolemy is determined by a combination of physical and mathematical considerations. Thus he locates the apex of the visual cone in the center of curvature of the cornea. As a result the visual bundle (whose behavior, as we noted above, is analogous to that of a bundle of light) will emerge unrefracted from the eye, thus saving the principle of the rectilinear propagation of the visual flux. Moreover, in Ptolemy's theory this point coincides with the center of rotation of the ocular globe. Thus he makes sure that in binocular vision the distance between the apexes of the two visual cones is invariant despite variability in the point of fixation.[1]

It is evident from all this that during the Hellenistic period optics had already developed into a comprehensive *theory of visual perception* in which physical, physiological and psychological questions were not treated as isolated problems which can be profitably studied by independent disciplines but rather as inextricably interlocking aspects of one and the same problem:

From here it is, in retrospect, a relatively small step to legitimize the use of visual rays as mathematical fictions representing the true geometrical properties of sight without attributing to them any physical (or material) reality. That step was taken by Alhazen. He regarded the rays as useful mathematical constructs, yet indispensable for a proper understanding of the visual process:

> "...all that mathematicians who hold the doctrine of the ray use in their reasonings and demonstrations are imaginary lines, which they call lines of the ray. And we have shown that the eye does not perceive any of the visible objects except through these lines. Thus the opinion of those who take the radial lines to be imaginary is correct, and we have shown that vision is not effected without them. But the opinion of those who think that something issues from the eye other than the imaginary lines is impossible, and we have shown its impossibility by the fact that it is not warranted by anything that exists, nor is there a reason for it nor an argument that supports it." [*Optics*, I.6; MS Fatih 3212, fol. 104a; tr. by A.I. Sabra (1978), pp. 180-1; also in the Latin version, known in the Middle Ages as *Perspectiva* or *De aspectibus*, in *Opticae Thesaurus. Alhazeni Arabis libri septem...*, [Basel (Risner), 1572; repr., New York, 1972], I, cap. 5, 23, p. 15, where these statements are somewhat compressed, according to A.I. Sabra; the Risner edition of Alhazen's *De aspectibus* will be referred to below as ASP].

Clearly, the problem Alhazen was struggling with, and which he solved in part, was the momentous problem of the methodological status of formal models and their applicability to empirical domains in spite of their lack of qualitative similarity to the realities they simulate.

1 Cf. A. Lejeune (1948), pp. 54-5.

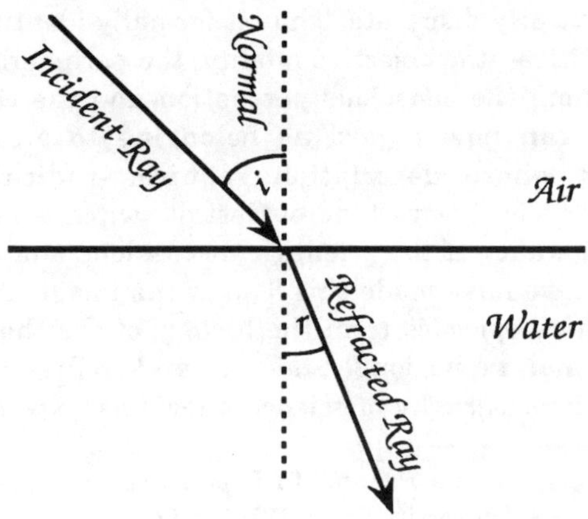

Fig. 6 — Modern theory of the refraction of light rays. *Light is bent when it passes obliquely from one medium to another due to differences in the speed of light in transparent media. Its angle with the normal will be less in the material in which it travels more slowly. Only light rays traveling along the normal will cross the boundary unrefracted. The law of refraction relates the angles of incidence and of refraction to their sines and to the relative velocities of light in the respective media. Its classical formulation was first worked out by the Dutch astronomer and mathematician Snell in 1621 in the formula* n *sin* i = n' *sin* r, *where* n *and* n' *are the indices of refraction of the relevant media.*

how, and what, does the living eye see and know?[1] This was the fundamental question which the various research traditions shared in common despite their vast differences in other respects. In conjunction with the epistemological postulate of the essential unity of knower and known as the condition of the possibility of perceptual knowledge as such this fundamental approach was to constitute the foundation of the identity theory of perception during the Arab and European Middle Ages, especially under the progressive influence of the basic ideas of Aristotelian thought. Even the weaker versions of this theory implied a negative heuristic which forged into one 'sympathetic

1 For a similar analysis cf. A.C. Crombie, 'The Mechanistic Hypothesis and the Scientific Study of Vision: Some Optical Ideas as a Background to the Invention of the Microscope', in *Historical Aspects of Microscopy, Papers read at a one-day Conference held by The Royal Microscopical Society at Oxford, 18 March, 1966,* S. Bradbury and G.L.E. Turner (eds.), (Cambridge, 1967), p. 12.

chain' elements and stages involved in the visual process which we cannot help but regard as utterly disparate. Thus it formally identified—and treated as homogeneous entities—the objective quality, the peripheral impression, the phenomenal sensation, the conscious perception and the certified cognition each of which we can now regard as belonging to altogether different ontological orders whose description requires radically heteronomic conceptual schemes.[1] Thus, some kind of Gestalt switch is necessary for us to understand the rationality of the scientific discussions and arguments of the time. However, once we have made this leap of the imagination, then quite a few apparently bizarre episodes from the history of the theory of perception will fall into place and we no longer need to seek refuge in that *refugium miserorum* of the historiography of science known as 'external history'.[2]

---

1 Cf. D. Davidson, 'Mental Events', in *Experience and Theory*, L. Foster and J.W. Swanson (eds.), (Mass. University Press, 1970), p. 94.

2 Cf. I. Lakatos (1971), pp. 91-135.

IV

THE IDENTITY POSTULATE AT WORK IN VARIOUS RESEARCH PROGRAMS IN THE THEORY OF VISION DURING LATE ANTIQUITY AND DURING THE ARAB AND EUROPEAN MIDDLE AGES.

1. The Identity Postulate at work in the Stoic-Galenic theory of vision.

Galen (2nd century A.D.) upon whose anatomical and physiological investigations the Arabs largely depended and whose theoretical insights thus exerted considerable influence, rejected the atomist theory of vision according to which material copies or *eidola* (compared by Lucretius to the cauls calves cast off at birth from the surface of their bodies[1]) penetrate the eye. The main reason for his rejection was the consideration that large *eidola* (e.g. of mountains) could not possibly enter through the small pupil of the eye and impress their real magnitude upon the sense of vision.[2] But this argument is relevant only if prior transformation (e.g., diminution) or else serial processing of the *eidola* must be regarded as barred by fundamental theoretical reasons. Clearly, the identity theory turns out to be the common presupposition.

Galen also rejected the Stoic analogy of the walking stick by means of which vision 'feels' the perceived object.[3] According to the Stoics the effluent visual pneuma 'pressurizes' the surrounding air in such a way that changes in pressure are immediately noticeable across indefinite distances. But in Galen's view this theory is entirely inadequate for the explanation of visual perception since the latter presupposes immediate consciousness. Conse-

1 Lucretius, *De rerum natura*, R.E. Latham (tr.), Penguin Books (1951), pp. 131-2.

2 *De placitis* 7, pp. 615,15—616,4 (K 5, 618).

3 *De placitis* 7, 642, 12-3 (K 5, 642). The Stoic model returns in Descartes' mechanistic analogy of the instantaneous propagation of light in a rigorously incompressible medium.

quently, the air itself must serve as a genuine "organ of discrimination for an object seen just as the nerves do for an object touched."[1] "But", the orthodox Stoic might object, "the nerve perceives nothing either. Rather it merely transmits the modifications caused by external objects to the *hegemonikon*, the true seat of consciousness." However, Galen thinks this psycho-physiological model entirely ill-conceived. No sensation of pain, he argues, could arise in any member of the body if that limb itself did not hold the faculty of sensation. Therefore, the true facts of the matter are as follows. The nerve is part of the brain. Thus, the entire member to which that part is attached fully receives the power of sensation and is capable of discriminating the things that touch it. Now an exactly analogous process enables the air under certain (pneumatic-luminous) conditions to discriminate visual impressions.[2]

Again, it is clear that Galen's argument heavily depends upon the postulate of the formal identity of perceiver, organ of perception and object perceived without which it is hardly intelligible. In this case it inspires the conception of vision as requiring some kind of organic psycho-physical continuum, a sympathetic chain of homogeneous elements stretching all the way from the seat of consciousness to the immediately sensed object.

2. The Identity Postulate at work in the geometrical tradition in the theory of vision.

Within the optical tradition building upon the work of Euclid, Hero and Ptolemy the identity theory seems to be least effective. Yet this is only apparently so. For even Alkindi's non-Aristotelian theory (9th century A.D.) implicitly presupposes (at least a weak version of) the identity theory. His most important argument designed to demonstrate the correctness of the extramission theory of visual perception adopted by him runs as follows. A paramount criterion of adequacy for any optical theory is that it explains *the primitive facts of visual perception*. These facts are always given in *perspective* (we do not normally see circles but only ellipses). Consequently, the intromission theory (of coherent forms) is untenable since it predicts that the eye, *and thus visual perception*, receives integral forms and qualities.

---

1 *De placitis* 7, p. 641, 11-2 (K 5, 641). Cf. S. Sambursky (1959), p. 127.

2 *De placitis* 7, pp. 641,13—642,13 (K 5, 641-2). Also cf. S. Sambursky (1959), p. 127.

Therefore, since all (in principle *viable*) versions of the intromission doctrine are hereby falsified, only the extramission theory of vision can be true.[1]

This argument makes no distinction between the geometrical, physical, physiological and psychological aspects or components of the visual process. In fact, in its positive as well as its negative aspect it is tied line, hook and sinker to the identity postulate. For the argument not only presupposes that the intromission and extramission theories are exhaustive theoretical alternatives (something Galen presupposed, too), but it also claims to effectively refute the idea of intromission as such. For these are clearly necessary (and jointly sufficient) conditions for the correct employment of the argument on behalf of the extramission theory. However, what the argument really refutes are no more than the *current* versions of the intromission theory, viz., those which hold that the transmission of forms involves wholesale transactions of integral true-to-life 'Gestalten' (or formally structured copies) rather than meaningless fragments or point-elements. This additional constraint, of course, is meant to select from the class of possible intromission theories only those thought to be viable in principle. Apparently this restrictive clause on the idea of intromission is taken as self-evident *by advocates and critics alike*. However, this restriction is far from obvious as would be shown much later. What, then, is the source of this universal consensus? I suggest that the implied reasoning is nothing but the identity theory being applied to the intromission hypothesis. This theory constituted, as it were, the common sense of every theoretician in the area of visual perception including even Alkindi and the extramissionists. Only a genius capable of conceiving a natural process guaranteeing the preservation of the original unity of the forms *despite* their fragmented entrance into the eye would be in a position to formulate a revolutionary theory of intromission which nonetheless would entirely fit within the constraints of the fundamental identity theory. That genius was Alhazen.

3. The Identity Postulate at work in Alhazen's theory of vision.

I cannot here discuss all the ingenious details of the synthesis achieved by Alhazen between the geometrical heritage of the perspectivists and the psychophysical potential of the Galenic-Peripatetic tradition. I am here primarily

1 *De aspectibus*, A.A. Björnbo and S. Vogl (eds.), Prop. 7, p. 9.

interested in uncovering the influence and the heuristic impact of the identity hypothesis. For a predominantly mathematical theory as that of Alhazen's this influence is surprisingly evident in quite a few respects.

The core ideas of Alhazen's theory are as follows. Already in Alkindi's *De Radiis Stellarum*[1] we find the fundamental notion which in the magico-mathematical tradition of the later Middle Ages would become very influential, viz., that everything in this world acts upon everything else by emitting specific radiation in all directions like a star.[2] The radiation of light is the prototype of this universal radiation. Consequently, Alkindi regards optics as the fundamental science.[3]

It is this idea which is seized upon by Alhazen. He modifies it in the sense that from every *point region* on the surface of bodies *elementary* forms (of light *and* color) are radiated in all directions along straight lines. In this way he meets the classical objection against the intromission theory that large (coherent) forms cannot enter into the eye.

But in other respects the new scheme of explanation appears to be utterly implausible. For how on earth is the coherence, the order and the integrity of the visual forms (*ṣūra*, εἶδος, *forma*) to be explained as a result of incoherent and chaotic sources of radiation? This challenging, but from the point of view of the traditional identity theory rather improbable, problem Alhazen solves by assuming that only those point forms perpendicularly incident upon the cornea, and thus continuing their journey unrefracted, will be effective in the process of visual perception. Thus, Alhazen has outlined the theoretical possibility of an exact homographic correspondence between the point forms of the object and the 'Gestalt' perceived. Consequently the formal requirement of the identity postulate has been satisfied in principle.

But the more detailed theoretical elaboration of Alhazen's theory of vision is also, and perhaps even more thoroughly, shaped by the identity postulate. In the first place the anatomical arrangement of the eye (for knowledge of which the Arabs depended mainly upon the work of Galen) is wholly dictated by the requirement of the geometrical transmission of point forms lest the

---

1 Digby 91, 16th century, fols. 66-80, *Alkindus de radiis stellarum;* alternative titles: *De radiis stellicis*, or *De radiis stellatis*; cf. Lynn Thorndike, *A History of Magic and Experimental Science*, (Columbia University Press, 1923), Vol. I, p. 643.

2 L. Thorndike (1923), p. 646.

3 Cf. Graziella Federici Vescovini, *Studi sulla prospettiva medievale*, (Turin, 1965), pp. 44-7; David C. Lindberg (1971), pp. 470-1.

necessary identity should get lost.[1] Thus, Alhazen takes all parts of the eye to be spherical and he supposes that all centers of curvature of the several refracting surfaces are situated upon a single line connecting the center of the pupil with the center of the optic nerve. By assuming that the centers of curvature of the cornea, the albugineous (or aqueous) humor and the anterior surface of the crystalline humor (or anterior *glacialis*) all coincide at a single point the theoretical requirement that all ocular surfaces through which light must pass before emerging from the posterior surface of the crystalline humor shall be concentric is anatomically satisfied. Consequently, all rays of light perpendicularly incident upon the eye will pass all refracting surfaces of the eye's optical system unrefracted. Therefore, a unique set of point forms identical to the visible object will reach the sensitive crystalline. This way the theoretical ideal of the classical theory of intromission is maintained: the living (or 'natural') eye receives intact the total visual form of the object seen.

This form should not be confused, however, with the modern notion of the real optical image. Despite the mechanistic appearance of some of Alhazen's theoretical schemes he is still a loyal captive of the conceptual framework associated with the identity theory. Thus, Alhazen does not describe a 'dead' mechanism of image formation in the eye as in a *camera obscura*. For in his view the faculty of insight inheres in the *living* eye alone. Only the living eye is sensitive to the forms of light and color from the visible object and perceives them *by assuming the corresponding qualities*.

This process first takes place, of course, in the seat of vision, or the glacial humor. Alhazen explains it as follows. The glacialis is somewhat transparent as well as dense. In this respect it resembles glass (*glacies*). Therefore, according to Alhazen, the forms may indeed penetrate the glacial humor on account of its *transparency*. But due to the *density* of the glacialis, which impedes their propagation, they cannot freely pass through it. Consequently, "the forms are fixed in its surface and body, albeit weakly."[2] Again, the same idea is clearly expressed at various places in book II of his *De aspectibus*. Thus he writes:

> Essential light (*lux essentialis*)...is perceived by the sentient body as a result of the illumination of the sentient body; and

1 Cf. Matthias Schramm, 'Zur Entwicklung der physiologischen Optik in der Arabischen Literatur', *Sudhoffs Archiv für Geschichte der Medizin und der Naturwissenschaften,* 43, 4 (1959), pp. 294-6. For the following account cf. 'De Compositione Oculi, Forma et Situ', ASP I, cap. 4, pp. 3-7.

2 ASP I, cap. 5, 25, p. 15.

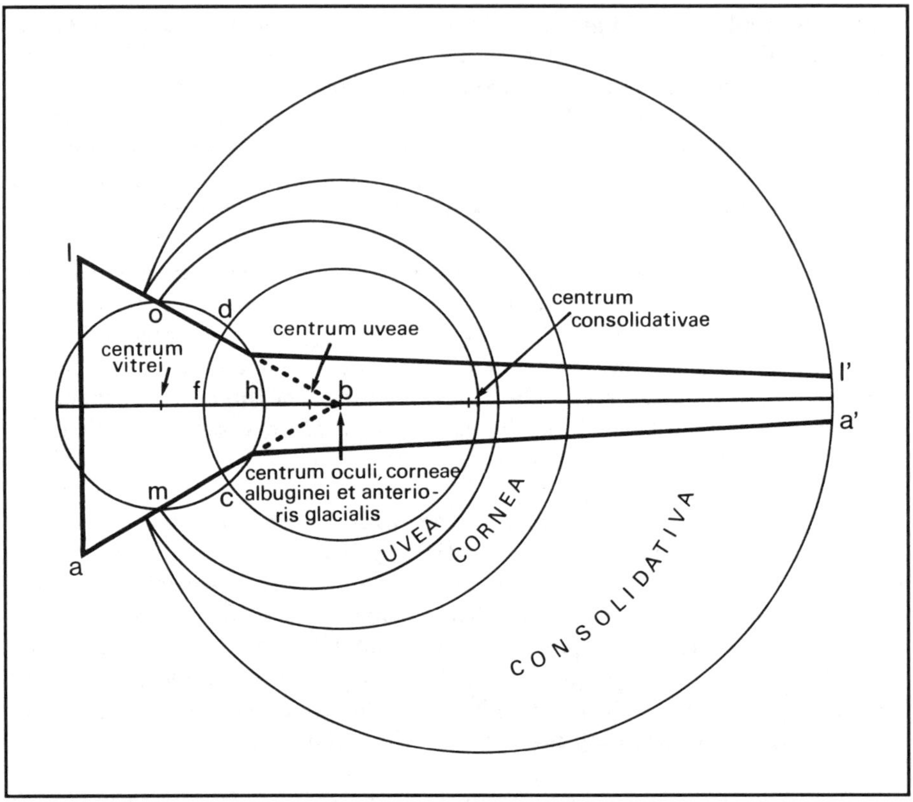

Fig. 7* — *Geometrical model of the eye *representing the various curvatures of the refracting ocular media according to Alhazen's geometrical optics of the eye.*[1] *The drawing is not intended to represent anatomical structures, of course, but only the trajectory of light as it passes through the transparent media of the eye. Accordingly, dots from left to right show* centers of curvature *of, respectively, the vitreous humor (i.e. of the concave* anterior *surface of the vitreous humor as it is hollowed out to receive the* crystallinus *or* anterior glacialis*; thus* centrum vitrei *here means* centrum concavitatis vitrei*); the* uvea *(which stands for the choroid, inclusive of the iris); the* cornea*, the* humor albugineus *(or aqueous humor) and the* anterior glacialis *(answering to what is now known as the lens of the eye), all three of these coinciding at the center of the eye; and finally the* consolidativa *(or sclera), the opaque fibrous tunic covering the greater part of the globe of the eye, and uniting in front with the transparent* cornea*).*

---

1 Slightly adapted from *The Opus Majus of Roger Bacon*, v, i, iii, John H. Bridges (ed.), (Oxford, 1897; Frankfurt/Main, 1964), ii, 24.

According to Alhazen, point forms (or 'rays') from the sighted object a-l *pass perpendicularly through the* cornea *at* m-o *and strike the anterior surface of the* glacialis *perpendicularly at* c-d, *thus penetrating unrefracted into the crystalline humor. However, at the interface between the* glacialis *and the vitreous humor they are being refracted away from the center of the eye so that they do not intersect but instead continue their journey as a single coherent whole along parallel lines, thus transmitting a formally and qualitatively identical representation of the object via the optic nerve to the* ultimum sentiens.

> color is perceived by the sentient body as a result of the alteration of the form of the sentient body and because of its coloration.[1]

Indeed, this old (Aristotelian) notion is also applied to the 'post-glacial' stages of the transmission of the forms of light and color so that the entire visual process can properly be described as a process of successive stages of continuous 'contagion':

> And the form of light and the form of color do not otherwise reach the concavity of the nerve except because the sentient body spread out in the concavity of the nerve is colored by the form of light and color and illuminated by the form of light. [Only then] the form is propagated to the concavity of the common nerve [i.e., the chiasma]. And any part of the sentient body in the concavity of the common nerve to which the form of the object seen is propagated will be colored with the color of that seen object and illuminated with the light in that seen object.[2]

Clearly, then, this entire process of continuous contagion by no means consists of (a chain of) merely mechanical operations. On the contrary, it presupposes a live eye capable of its natural sensitive functions. To be sure, Alhazen compares the fixation of forms in the glacialis to the fixation of light in a transparent but dense body. But this is no more than a harmless analogy. Indeed, Alhazen adds, quite significantly:

---

1 ASP II, cap. 2, 18, p. 35. Also cf. ibid., II, cap. 2, 19, p. 36: "... when the form of light and color passes through the eye, the eye is colored, and when the eye is colored, it perceives that it is colored, and thus it perceives the color."

2 ASP II, cap. 2, 16, pp. 34-5.

> And the glacialis is also endowed with the power to receive these forms and to perceive them. Thus it is [only] in virtue of its sensitive power to receive that the forms permeate it.[1]

Thus one may even be inclined to pass Alhazen's theory off as merely good old Aristotelian stuff. But that would be equally unjustified. To be sure, Aristotle provided the general categories and the scheme of explanation employed by Alhazen. But compared to the richness and the concreteness, to the theoretical rigor and the comprehensiveness of the latter optico-psycho-physiological account Aristotle's theory is no more than a bleak metaphor.[2] Especially in his account of the details and the mode of interaction between the 'passive' glacialis and the actions of light and color Alhazen goes far beyond Aristotle. Thus he continues the above passage as follows:

> And when the form reaches the surface of the glacialis, it acts on it, and as a result the glacialis suffers. For it is in the nature of light to act upon the eye and it is in the nature of the eye to suffer under the action of light. And this effect, which light produces in the glacialis, is propagated through the glacial body only along the rectitude of the radial lines, because the glacialis is endowed with the power to receive the forms of light from the vertices of the radial lines... And it is due to this action and [the resulting] pain that the glacialis has a sense of the *forms* of the visible things on its surface and throughout its entire body. And it is due to the order of the parts of the form on its surface and throughout its entire body that it has a sense of the *order* of the parts of the agent.[3]

In this passage the Aristotelian physiology of vision is not only enriched with the view that visual sensation is a species of pain.[4] But it is also combined with the doctrine, referred to above, that the parts of the form projected onto the eye must isomorphically preserve the order of the parts of the visible object from which that form originated. And we will see that this novel element furnishes the cornerstone of a most powerful and fertile synthesis in the theory of vision in which the former incompatibility between the two theoretical traditions now miraculously forged together by Alhazen's genius (viz., the Aristotelian vs. the geometrical tradition) seems to have been '*aufgehoben*' in truly Hegelian fashion to give way to an entirely new research program.

1 ASP I, cap. 5, 25, p. 15.
2 Cf. *De anima* 2, 12, 424a 17-23.
3 Loc. cit.; emphasis added.
4 Cf. D.C. Lindberg (1976), p. 71; A.I. Sabra (1978), pp. 166-8.

However, before going deeper into the intricate arcana of Alhazen's theory of vision I must mention two further details of his anatomical construction.

Alhazen also assumed that the common center of curvature of the various refracting surfaces exactly coincided with the center of the eye. Consequently, the position of this point becomes independent of the various axial rotations of the eye. A similar geometrical ingenuity of ocular anatomy, dictated entirely by theoretical considerations, could already be found in Ptolemy.[1]

Finally Alhazen's anatomical construction guarantees that the radial lines of the eye (the so-called 'lines of the (visual) ray' in the extramission theory) are identical with the perpendiculars on the ocular surface. Thus Alhazen writes:

> The nature of sight is to receive what comes to it of the light of visible objects, and...its nature is further characterized by receiving only those forms that come to it through certain lines..., namely, the straight lines whose extremities meet only at the center of the eye, these lines being alone characterized as diameters of the eye and perpendicular to the surface of the sensitive body [i.e. the crystalline humor]. Thus perception occurs through the forms coming from the visible objects, and these lines are, as it were, an instrument of sight by means of which the visible objects appear to it distinct and the parts of each visible object ordered.[2]

But how does Alhazen propose to explain the required selective effectiveness in the perceptual process of the radial forms alone? In other words, how does the filtering of 'incident' forms take place as required by the theory? We will find that in Alhazen's expositions concerning this vital obstacle in his scheme of explanation the identity theory is operative even more strongly than in the part discussed above, especially in the implied requirement (one might call it the weak version of the identity theory) that the theory of vision should explain how (under standard conditions) the *living* eye is *immediately* capable of *veridical* perception of concrete *natural* objects.

---

1 Cf. A. Lejeune (1948), pp. 54-5. Also cf. above p. 34. It is interesting to compare with this theory the modern 19th century controversy surrounding the theory of the *anatomical* identity of the so-called corresponding points on the sensitive surfaces of the two retinas (cf. ch. IX, sec. 3-4).

2 *Optics* I, 6; MS Fatih 3212, fol. 97b; tr. by A.I. Sabra (1978), p. 181; ASP I, cap. 5, 20, pp. 12-3.

In the first place we have already noted that on Alhazen's account the eye is not a purely optical mechanism but a sensitive instrument of perception (and even of cognition). Galen had already singled out the crystalline humor (the *crystallinus* or *glacialis*) as the sensitive organ of the eye. In Alhazen this aspect acquires central theoretical significance. For him it means that the reception of the forms (of light and color) cannot be explained purely 'mechanically' but rather as determined by what he regards as the essentially *dual* nature of the glacialis. For on the one hand the glacialis is no more than a transparent substance (just as air and other non-sensitive bodies). But on the other hand the glacialis is at the same time a sensitive organ. And this should not be understood in a metaphorical sense like when we speak of a sensitive instrument. For Alhazen the sensitivity of the glacialis has only one, and that an expressly non-mechanicist, sense: here, in this transparent body, *sensation* (*sensus*) occurs. And this sensation is propagated from the eye to the common nerve just as the sensations of pain and touch are propagated. And it is from the sensation produced in the common nerve that the sensitive faculty perceives the form of the visible object.[1]

---

1 Cf. ASP I, cap. 5, 27, p. 16; also cf. ibid., 25, p. 15, quoted above. D.C. Lindberg charges that Alhazen equivocates on the issue of whether the sensation is propagated to the *ultimum sentiens* all by itself or whether it is still accompanied by the form that produced it upon its contact with the glacialis (cf. his (1976), pp. 83-4). However, this alleged equivocation may be due not to Alhazen himself but rather to the Latin translation of his text as it appears in the Risner edition which Lindberg uses as his source. By contrast, the relevant passages translated by A.I. Sabra from Arabic MSS certainly provide at least partial clarification (cf. A.I. Sabra (1978), pp. 167-8).

Nevertheless, Lindberg's generalized comment concerning Alhazen's "basic indecision over the relative functions of the various sensitive elements in the visual pathway" (loc. cit., p. 84) also points to a more fundamental ambiguity concerning the relative contributions of psychological vs. physiological factors and their mutual relations in the visual process. Of this ambiguity it is certainly right to say, as Lindberg does, that it "would remain until the seventeenth century" (ibid.). However, pursuing a deeper, and hopefully more illuminating level of historical analysis we had better refer this wide-spread and pervasive ambiguity not so much to personal indecisions on the part of individual scientists but rather to the tensions generated by a comprehensive research program into the theory of vision whose negative heuristic *forbade* splitting up the visual process into separate empirical and philosophical domains that can be profitably studied in isolation from each other (cf. above, ch. III, sec. 4). As a result, medieval theories of vision failed to make a sharp ontological as well as a methodological distinction between the psychological and the physiological aspects of vision. Or rather, given their powerful research program it would be absurd for them to deny the formal unity of the visual impression and the corresponding sensation. Just as absurd as it

As a result of this double nature of the glacialis there are also two modes of reception of the forms: one in virtue of its sensitive nature and one in virtue of its transparency.

> The sensitive organ [i.e., the crystalline] does not receive the forms in the same way as they are received by transparent bodies. For the sensitive organ receives these forms and senses them, and the forms go through it on account of its transparency and on account of the sensitive power that is in it. Therefore, it receives these forms in the manner proper to sensation, whereas transparent bodies receive them only in the manner proper to transmission without sensing them. And if the sensitive body's reception of these forms is not like their reception by non-sensitive transparent bodies, then the forms do not extend through the sensitive body along the lines required by sensitive bodies, but rather along the extension of the parts of the sensitive body. Sight is thus characterized by receiving the forms along the radial lines alone, because it is a property of forms to extend in transparent bodies along all straight lines and therefore they come to the eye along all straight lines. But if sight received them along all lines on which they arrive, the forms would not [appear] to it ordered. And therefore sight has come to be characterized by receiving the forms through those [radial] lines alone, so that it would perceive the forms with the order they have on the surface of visible objects.[1]

Thus the physical or inorganic mode of reception only governs the transmission of any form reaching the surface of the glacialis along any straight line. However, this process alone would radically distort the original order of the point forms on the visible object. It is only in virtue of the co-activity of an auxiliary process of reception characteristic of the *sensitive* nature of the glacialis that the original order is *selectively* preserved. This selectivity, in other words, is not the simple result of any superior force attributable to the forms entering through radial lines (though this characteristic may help in sorting them out[2]). Rather it is a consequence of the distinctively *psychological* power of discrimination with which the eye as a living organ of

---

would be, say, for the Cartesians a few centuries later to suppose that there was more to mature perception than immediate sensation.

1 *Optics*, II, 2; MS Fatih 3213, fol. 7a-b; tr. A.I. Sabra (1978), pp. 165-6; ASP II, cap. 1, 4, p. 26.

2 Cf. *Optics* I, 6; MS Fatih 3212, fol. 90b; tr. A.I. Sabra (1978), p. 182; ASP I, cap. 5, 18, p. 10, lines 27-30.

man's sensory apparatus is endowed by nature and in virtue of which it is radically distinct from a mere mechanical optical instrument. This is not to deny that Alhazen offered mechanical analogies to account for the superior strength of perpendicular rays.[1] But on the whole Alhazen's ingenious theorizing remained squarely within the framework defined by traditional epistemology in terms of which a radically mechanical solution of the problem of vision was bound to be inconceivable. Thus the fundamental difference between Kepler and Alhazen is precisely the latter's insistence on a solution of the problem of non-perpendicular rays in terms of selective sensitivity, i.e., his insistence on a *psychological* rather than an exclusively *mechanical* solution. This is entirely in accordance with the identity theory according to which what optics studies is the *living* (*'natural'*) eye and not a dead optical instrument. By contrast, Kepler's theory of the crystalline as a lens rather than as a sensitive body presupposes a theoretical framework in which the eye, dead or alive, is studied as no more than a mechanico-optical instrument capable of variable accommodation and of relatively sharp image formation.[2] Alhazen's notion of selective sensitivity, however, is *psychologically* explained as a more or less intentional function of the living glacialis with which it is endowed in virtue of its sensory character. *Consequently, the two theories also entail contradictory predictions*. For even if one regards the Alhazenian forms in a non-Aristotelian sense as (point) images, still Alhazen's theory, unlike Kepler's, would not predict the formation of a real optical image behind the transparent substance of a *dead* eye, but only (or at best) a weak diffuse light. Thus Scheiner's (1625) and Descartes' (1637) experiments[3] whose results accorded with the implications of Kepler's theory (by producing real inverted images on the retinas of freshly dead animal and human eyes) could be taken as straightforward corroborations of the mechanicist hypothesis and as 'undeniable' testimony underscoring the degenerateness of the identity theory of perception. And so they were.

From a modern point of view, therefore, the glacialis in Alhazen's theory is promoted into an unlikely psycho-physical hybrid simultaneously governed by quasi-intentional and quasi-mechanical laws which *jointly* determine, as vectors in one and the same field as it were, a suitable response. However, this train of thought becomes much more intelligible once we see it as a consequence of fundamental theoretical postulates. The epistemological re-

---

1 Cf. e.g. ASP VII, cap. 2, 8, p. 241. Also cf. D.C. Lindberg, 'The Cause of Refraction in Medieval Optics', *Brit. J. Hist. Sci.* 4, (1968-69), pp. 25-9.

2 Cf. A.C. Crombie (1967), p. 54.

3 Cf. figure 15, p. 81.

quirement of the formal unity of (what we would now radically differentiate as) the distal focus, the proximal stimulus, the peripheral and central responses spontaneously suggested the demarcation of an excessively comprehensive domain for optics. The resulting 'optics' was both anthropocentric and naturalistic. It was neither a pure theory of light, nor a pure theory of vision. Rather, 'optics' was to be a comprehensive theory of the 'natural' (i.e., organic rather than mechanical) transactions in visual perception in which physical and perceptual questions could be treated as mutually dependent and as theoretically solvable within a common framework.

This interpenetration of the various (optical, physiological and perceptual) aspects of visual perception is also strikingly manifest in Alhazen's quasi-optical treatment of the processes transmitting the perceived forms to the *ultimum sentiens* where visual perception reaches its 'completion'.[1] However, I must first indicate just how the epistemological postulate of the formal identity of subject and object of perception shapes the heuristic of Alhazen's theory formation. For why is it that visual perception already begins at the glacialis? The answer must be: because it is there, and only there, that the forms are arranged in a way which guarantees the formal identity of perceiver and perceived, thereby creating the possibility of *veridical* perception. But—one is inclined to ask—why is it necessary for this formal arrangement to take place here and not at some later stage in the visual process? The answer to this question is, *from the perspective of the identity theory*, absolutely obvious. For if the 'radial' forms, on which the reliability—or, indeed, the very possibility—of visual perception depends, should not be selected in virtue of the *sensitive* nature of the glacialis, then their further transmission would have to be governed exclusively by the *transparent* nature of the ocular substances. Consequently, they would all coincide in one single point, viz., the center of the eye. But if their journey should terminate there, then, reduced to a single point, they would be altogether incapable of communicating to the *ultimum sentiens* the order of the parts of the visible object. On the other hand, in case they would continue their journey beyond the center of the eye they would completely distort the proper order of the parts of the object and transform it in its (doubly) symmetrical opposite. In both cases, therefore, the formal identity of the essential qualities of the perceived object and the corresponding visual in-form-ation which was supposed to be necessary for veridical visual perception would be fatally flawed.[2]

1 "...sensation is not fully accomplished by the eye alone, but only by the *ultimum sentiens.*" (ASP I, cap. 5, 27, p. 16, lines 31-2).

2 ASP II, cap. 1, 2, p. 25.

In order to avoid this intolerable violation of the identity postulate, therefore, Alhazen assumes (and this is again an anatomical assumption apparently dictated by purely theoretical considerations) that the interface between the glacialis and the vitreous humor lying behind it (1) is located in front of the center of the ocular globe and (2) serves as a *refracting* surface for the (ordered and registered) forms. As a result of this refraction the selected point forms are no longer propagated along the radial lines but are deflected before they reach the center of the eye and continue their journey to the *ultimum sentiens* along parallel lines, i.e., as a single coherent whole.[1]

However, supposing the radial forms had not been selected at the surface of the glacialis, then the question would arise how they could possibly be selected at any later stage of the visual process *after they had undergone refraction* (just as all other 'non-radial' forms), thus *losing their sole distinctive character*. For, surely, on Alhazen's account the selective power of the glacialis can be satisfactorily explained not only in terms of its presumed exclusive sensitivity along certain preferred directions defined by the radial lines (as though it consisted of numerous little tubes stretched out along the radial lines whose opaque walls would absorb any ray of light not exactly incident along the rectitude of these lines[2]). But this merely *postulated* directional sensitivity can further be justified by reference to the superior strength of action of the forms incident along the radial lines, as all and only such forms perpendicularly incident upon the glacialis can penetrate it without being refracted.[3] Thus, the selective potential of the *glacialis* can in principle be explained as a result of its sensitivity being properly attuned to the superior impact characteristic of unrefracted forms only.

---

1 ASP II, cap. 1, 2, p. 25.

2 Such a model is vaguely suggested by such locutions as:

> "... the forms do not extend through the sensitive body along the lines required by transparent bodies, *but rather along the extension of the parts of the sensitive body*. Sight is thus characterized by receiving the forms along the radial lines alone." [*Optics*, II, 2; MS Fatih 3213, fol. 7a-b; tr. A.I. Sabra (1978), p. 166; ASP II, cap. 1, 4, p. 26; emphasis added].

3 *Optics*, I, 6; MS Fatih 3212, fol. 90b; tr. A.I. Sabra (1978), p. 182; ASP I, cap. 5, 18, p. 10, lines 27-30:

> "But the effect of the lights that come along the perpendicular is stronger than the effect of those that come along inclined lines. Therefore, it is most appropriate that the crystalline should perceive, through each point on it, the form that comes to this point along the perpendicular alone, without perceiving through the same point that which comes to it along refracted lines."

However, for the *vitreous humor* to perform a similar feat of selective sensitivity after the (originally) 'radial' forms would have been *refracted* at the posterior surface of the glacialis would require either an arbitrary *directional* sensitivity (along lines parallel to the line of ocular centers) that would go entirely unexplained as it could not be based upon any optical feature uniquely characteristic of the selected subset of incident forms. Or else one would have to postulate a selective sensitivity to certain *angles of incidence*, each one of which would have to be different for each point of the interface. Needless to say, both theoretical alternatives would be as implausible as they are *ad hoc*. Thus the ineluctable fact remained that on the one hand refraction of the radial forms was required in order to maintain the formal identity dictated by the identity postulate. But on the other hand no plausible retrieval of the radial forms seemed possible after their necessary refraction had actually taken place.

This complex line of reasoning in which the identity postulate plays a central role must have seemed obvious to Alhazen. Consequently, he must have felt forced by reason rather than by mere tradition to abide by the old Galenian doctrine that the crystalline humor is the first and most excellent sensitive organ in the visual pathway. And he must have felt confident in ascribing to its established sensitivity the novel power of arranging the incident point forms which guaranteed the essential identity of perceiver and perceived.

However, this analysis also necessitated in Alhazen's theory a remarkable quasi-optical treatment of the psycho-physical process which conducts the sensation formed in the glacialis via the vitreous humor and the *opticus* to the optic chiasma, where, in unison with the corresponding form from the other eye, it 'in-forms' the *ultimum sentiens*. For it is noteworthy to observe that according to Alhazen the refraction at the interface between the crystalline and the vitreous humor is co-determined by the double nature of the ocular substances involved. Thus the reality of the co-activity of the relevant psychophysical processes is not an *ad hoc* assumption invoked only when it is convenient (as one might suppose in the case of the explanation of the postulated selective power of the glacialis). For in this case the assumption creates an unexpected difficulty which Alhazen describes in meticulous detail:

> Therefore, forms are refracted at the vitreous humor as a result of two causes, one of which is the difference of the relative transparencies of the two bodies [i.e., of the vitreous humor and the (anterior) glacialis], the other the difference of their relative sensitivities. And if [only] the transparencies of these two bodies were the same [while the sensitivities differed], then the form would be transmitted in the vitreous body along the

> rectitude of radial lines because of the similarity of their transparencies, while at the same time it would be refracted because of the difference in their relative sensitivities. And thus, because of the refraction, the form would be monstrous, or [rather] it would be two forms because of this discrepancy.[1]

In other words, because there are two causes of refraction the refractive power of these two 'natural' operations must be exactly the same if vision is to be possible.

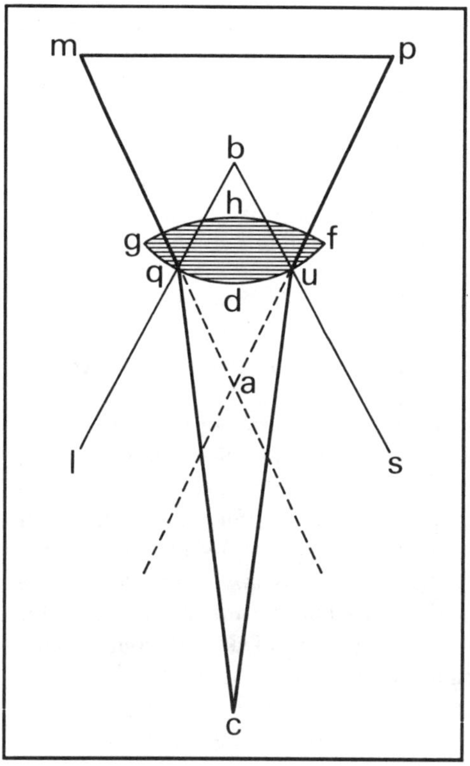

Fig. 8 — Radiation of point forms ('rays') *through the glacial humor according to Roger Bacon following Alhazen.*[2] *The visual pyramid has the object sighted* m-p *as its base. Rays strike the anterior surface* g-h-f *of the glacialis (or crystalline 'lens') perpendicularly and thus pass through unrefracted, directed toward an apex* a *at the center of the eye, which is also the center of curvature of the (anterior) glacialis. However, before they intersect at* a *they are refracted away at* q *and* u *on the posterior surface of the glacialis, thus projecting an erect 'image' to an apex at* c *in the common nerve. Lines* b-q-l *and* b-u-s *are are the normals on the points of incidence* q, u. *Thus* b *represents the center of curvature of (the concave anterior surface of) the vitreous humor. However, Bacon (and Alhazen) are mistaken about the refractive properties of the ocular media in question, as the refractive index of the vitreous humor is in fact* lower *than that of the (anterior) glacialis.*

1 ASP II, cap. 1, 5, p. 25, lines 5-10.

2 Slightly adapted from *The Opus Majus of Roger Bacon*, v, i, vii, John H. Bridges (ed.), (Oxford, 1897; Frankfurt/Main, 1964) ii, 48.

In the third place the quasi-optical character of the psycho-physical process becomes apparent in Alhazen's theory of the *non-rectilinear* propagation of the form (of light!) possibly already in the vitreous humor but at any rate in the optic nerve. This deviation from rectilinearity is dictated by the theoretical necessity of an exact point-to-point reunion of the forms arriving from the two eyes in the optical chiasma. As a result the *ultimum sentiens* perceives one and the same form which represents an exact qualitative copy of the perceived object.[1]

The profound extent to which the identity postulate defines, for Alhazen, the entire heuristic matrix for any theory formation concerning visual perception is also evident, finally, from the fact that even epistemological distinctions are defined in purely optical terms. This is clear from Alhazen's doctrine of 'ascertained' forms (*forma certificata*[2]). Not all vision is equally clear. Indeed, there are two modes of perception: superficial perception (*comprehensio superficialis* or *comprehensio per aspectum*) and directed perception (or *comprehensio per intuitionem*).[3] Only the latter deploys the central ray along the axis of the visual pyramid. For that is the only ray which also passes unrefracted through the interface between the crystalline and the vitreous humor and which is therefore stronger than all other rays in accordance with the principle that refraction always weakens.[4] A superficial impression is corrected and verified when the central ray, due to a rapid movement, sweeps across the entire visible object, thereby enabling the various visible qualities to reach the center of visual perception in uninterrupted succession without any qualitative or quantitative impoverishment. Only this kind of visual perception *(per intuitionem)* guarantees the reliability of the received form and even *precludes any possibility of error*[5]: for it alone certifies the reception of the true form (*forma vera*) in which all visible qualities of the perceived object are manifest. Here, in this doctrine, we touch upon the advanced psychological elements in Alhazen's theory of perception. Unfortunately, these cannot be discussed in this context.[6]

---

1 ASP, II, cap. 1, 5-6, pp. 26-7.

2 Cf. ASP II, cap. 3, pp. 67 ff; esp. sec. 64, p. 67, lines 20-5; or sec. 66, p. 69, lines 1-4.

3 ASP II, cap. 3, 64, p. 67, lines 20-2.

4 ASP II, cap. 1, 8, p. 29.

5 ASP II, cap. 1, 8-9, pp. 29-30.

6 For Alhazen's psychology of visual perception cf. Hans Bauer, 'Die Psychologie Alhazens auf Grund von Alhazens Optik dargestellt', in *Beiträge zur Geschichte der Philosophie des Mittelalters*, Vol. X, no. 5, (Münster, 1911).

4. The Identity postulate reinforced by the Baconian-Alhazenian synthesis in optical theory. Internal explanations facilitated by the proposed rational reconstruction.

The above analyses clearly show how incisive and fundamental the methodological and heuristic role of the epistemological postulate of identity has been in the various optical traditions culminating in the work of the greatest of medieval opticians, Alhazen. This role was only to become more pronounced during the Middle Ages in Western Europe. The theories of perception put forward by the later perspectivists in the tradition of Roger Bacon (such as John Peckam, Witelo, Henry of Langenstein, Blasius of Parma) remained on the one hand within the circle of influence and inspiration defined by Alhazen's theory of vision, on the other hand they were more and more influenced theoretically by certain Scholastic-Aristotelian conceptions which were almost universally adopted. In part this was a consequence of the optical synthesis achieved by Roger Bacon which, born from the desire to demonstrate the unity of knowledge, tried to reconcile Alhazen's work with fundamental tenets from Augustinian and Aristotelian traditions. Thus, Bacon identified Alhazen's notion of form (*ṣūra, forma*) with the Scholastic notion of *species*.[1] And in his Neo-Platonic theory, presumably borrowed from Robert Grosseteste, of the universal *multiplicatio specierum* (of which the propagation of light was the paradigm) he emphasized that what was involved here was not material emanations but only a process by which formal similitudes or species were communicated via one or more media, and that by means of qualitative changes preserving the specific identity of their efficient cause.

> Every efficient cause acts through its own power, which it exercises on the adjacent matter, as the light [*lux*] of the sun exercises its power on the air (which power is light [*lumen*] diffused through the whole world from the solar light [*lux*]). And this power is called 'likeness', 'image', and 'species' and is designated by many other names, and is produced both by substance and by accident, spiritual and corporeal... This species produces every action in the world, for it acts on sense, on the intellect, and on all matter of the world for the generation of things...[2]

---

1 *De multiplicatione specierum* 1, 1, in *The Opus Majus of Roger Bacon*, J.H. Bridges (ed.), 2 vols. (Frankfurt/Main, 1964), 2, 407-10.

2 *Opus Majus*, 4, 2, cap. 1, J.H. Bridges (ed.) 1, p. 111; tr. in E. Grant (ed.), *A Source Book in Medieval Science*, (Cambridge, 1974), p. 393.

Bacon, then, clearly adopted the Alhazenian intromission scheme in conjunction with an Aristotelian epistemology according to which vision was the result of the passive reception by sense of the species of the sensible thing.[1]

These and similar developments could only serve to reinforce the grip of the apparently fundamental identity postulate on the theoretical foundations of 'optics', even on those of the mathematical perspectivists. This interpretation may shed new light on the curious fact that certain developments in optical technology were systematically ignored, or dismissed as irrelevant, by the theoreticians of the later Middle Ages. During the 13th century glassworkers discovered the technique to grind correction lenses for presbyopia. Soon thereafter correction lenses for myopia were developed as well. But the academic world largely ignored the potential theoretical significance of these techniques. Indeed, this strange phenomenon was not an isolated one. Other strictly technical knowledge for practical use available among craftsmen, artists and engineers was equally neglected by the learned Academy. But was this attitude really as curious as it is sometimes made out to be?

Modern historiography of science all too easily seeks refuge in facile 'externalistic' explanations.[2] Thus popular theories have it that an unbridgeable social gap existed between the scholarly academicians and the formally uneducated craftsmen whose technical genius stemmed from manual, not from intellectual labor, and was therefore deemed to lack intellectual relevance.[3] But against the background of the fundamental postulate of the identity theory an internalistic explanation is much more plausible. For then it becomes clear that from a theoretical point of view any optical instrument could only distort the natural reception of forms and break the organic unity of subject and object. For natural (unimpaired) vision sees reality *essentially as it is in itself*. But eye glasses cannot but destroy this essential harmony. They distort or discolor[4], magnify or diminish[5], remove or bring

---

1 *De scientia perspectiva* 1, 5, cap. 1; *Opus Majus*, 5, J.H. Bridges (ed.) 2, pp. 30-2.

2 Cf. I. Lakatos (1971), pp. 91-135.

3 Cf. Edgar Zilsel, 'The Origins of William Gilbert's Scientific Method', *J. Hist. Ideas* 2 (1941), pp. 1-32.

4 Witelo and Theodoric of Freiberg had studied the colored forms projected onto a screen by light refracted by a crystal. Cf. A.C. Crombie, *Robert Grosseteste and the Origins of Experimental Science 1100-1700*, (Oxford, 1953, 1962), pp. 216-8, 230-1 (Witelo), 248-50 (Theodoric of Freiberg).

5 Such phenomena had been studied, i.a., by Grosseteste and by Roger Bacon. Cf. E. Rosen, 'The Invention of Eyeglasses', *J. Hist. Med.* XI, (1956), pp. 13-53, 183-218.

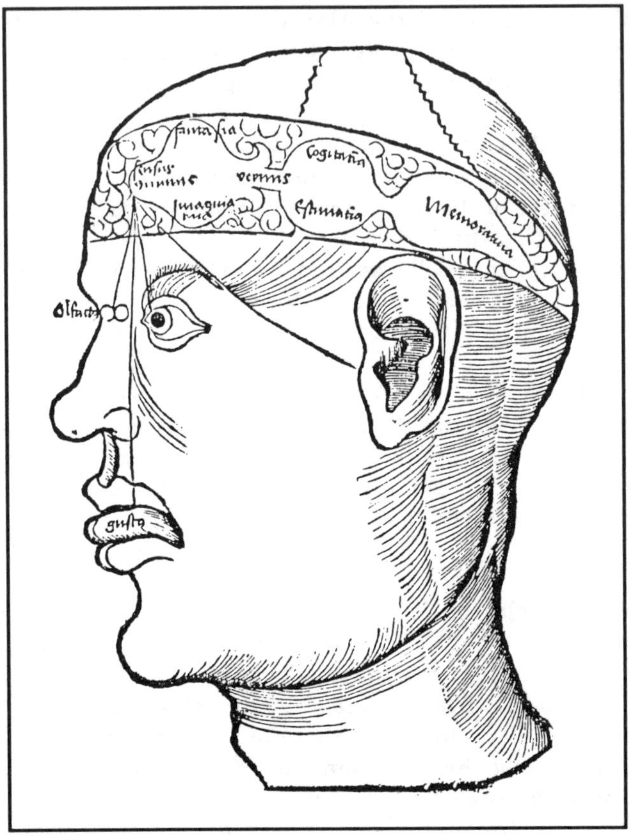

Fig. 9 — Traditional early theory of the 'functional architecture' of the brain.[1] *According to this scheme, based upon Galen and Avicenna, the frontal ventricle or cavity of the brain was the seat of the* sensus communis *or* imprensiva, *(i.e. the sensory center connected by (hollow!) nerves to the various sense organs), together with the* fantasia *and the* imaginatio. *The middle ventricle was occupied by* logitatio *and* estimatio, *i.e. the seat of thought and judgment; while the posterior ventricle was the seat of memory.*

It is interesting to note that Leonardo later modified this scheme and located the common sense in the second ventricle, the seat of the cognitive faculties. This theoretical adjustment may have resulted from his newly acquired appreciation of the cognitive qualities of sight. The mechanicist tendency, already implicit in his treatment of the eye as a camera obscura, *inclined him to reject the traditional doctrine of the crystalline humor as the seat of vision. He*

1 Illustration taken from Gregor Reisch, *Margarita Philosophica*, (Strasbourg, 1504).

emphasized that vision proper occurs in the common sense alone. Thus he may have anticipated the view that vision, or perception in general, is in fact a species of judgment.

---

closer. Instead of the essence of things optical apparatus only produced—could only produce—mere appearances. These county-fair gadgets, therefore, might conceivably have some practical utility, but from an epistemological point of view their effect was necessarily deceptive. Consequently, there was no theoretically adequate ground for their deployment in the context of serious research into the theory of vision. Surely, the *social gap* between academicians and craftsmen may have facilitated this theoretical dismissal, nonetheless it does not explain it. For an adequate understanding we have to become more aware of the *cognitive gap* separating medieval theory formation concerning vision from that of the 17th century. An alternative mechanistically inspired theory opening up the possibility of interpreting eye glasses as *correction* glasses for the impaired lens *mechanism* of the eye; or providing a theoretical framework within which the human body as a whole could be regarded as a physiological mechanism whose sensory functions can simply be extended by optical instruments; that kind of theory could not possibly be incorporated into the Scholastic framework without radical changes in the intellectual foundations of the medieval world of thought.

5. The internal disintegration of the research program defined by the Identity Postulate during the sixteenth century.

However closely medieval opticians apparently approximated the eventual solution of the dioptrical mechanism of the eye, yet until the very end of the 16th century the identity theory continued to suggest problems and solutions which were basically incompatible with the mechanistic methodology. The latter was to investigate physical and physiological questions independently of the psychology and epistemology of perception. Therefore, the *camera obscura* model of the eye would become its favorite model barring all non-mechanical questions now considered to be irrelevant. In the remaining part of this section I want to pursue the theoretical developments during the century prior to Kepler in order to show how the identity postulate gradually eroded under the impact of the very theories it inspired.

As far as we know Leonardo da Vinci was the first to suggest the *camera obscura* model of the eye (possibly followed by Maurolico of Messina and at any rate the Neapolitan 'magician' Giambattista della Porta). Indeed, Leonardo's methodology sometimes seems to be based upon the conception of the eye as a purely physical system. Thus he writes a note entitled: "How to make an experiment which demonstrates how the visual virtue employs the instrument of the eye."[1] Occasionally he also presents mechanical solutions for the dioptrical structures of the eye (e.g., the pupil as the place where incident light rays intersect; or the crystalline as a pure instrument for the refraction of light).

Nevertheless, he fails for all that. Not only because of his poor understanding of the rich theoretical tradition before him and of the central problems it attempted to solve (especially the problem of the arrangement of the non-selected 'non-radial' light rays reaching the eye from every point in the field of view). But also, and primarily, because of the identity theory. Leonardo, no more than the great perspectivists of the 16th century, managed to break away from the epistemological requirement that the eye mechanism constitutes the immediate cause of perception by delivering to the *opticus* or the *imprensiva* (the faculty of judgement) an image that is essentially identical with the visible qualities of the object perceived. It is this negative heuristic which is responsible for Leonardo's construal of his central theoretical problem: the inversion of the optical image.

In Alhazen's construction this inversion (deemed to be absurd) could take place only if the radial forms could pass through the center of the eye. But due to Leonardo's *camera obscura* model intersection of incident rays already occurs in the pupil. Consequently Leonardo assumes the *necessity* of a second point of intersection: "Necessity has provided that all the images of objects in front of the eye shall intersect at two places."[2] As regards the exact location of this intersection Leonardo offers conflicting theories. Sometimes, as in the passage quoted, he seeks this (second) intersection in the crystalline. Elsewhere[3] the second intersection occurs between the crystalline and the optic nerve.

---

1 Leonardo da Vinci, *Of the Eye*, in D.S. Strong, *Leonardo da Vinci on the Eye, The MS D... Translated into English and Annotated with a Study of Leonardo's Theories of Optics*, (Ph. D. diss., UCLA, 1967), p. 53; also in 'Leonardo da Vinci: Of the Eye', Nino Ferrero (tr.), *Am. J. Ophthalm.*, ser. 3, 35 (1952), p. 509.

2 Leonardo da Vinci, *Literary Works*, J.P. Richter and I.A. Richter (eds./trs.), (Oxford, 1939), I, 144. Also in I.A. Richter (ed./tr.), *Selections from the Notebooks of Leonardo da Vinci*, (Oxford, 1952), p. 116.

3 D.S. Strong (1967), p. 56; N. Ferrero (1952), p. 510.

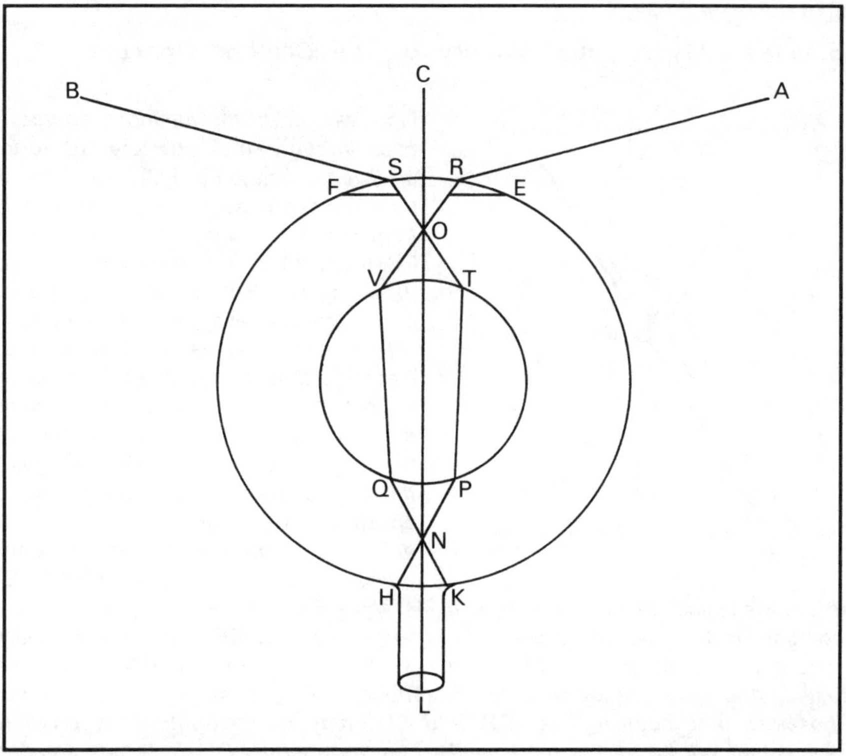

Fig. 10 — One of Leonardo's schemes for the propagation of light through the eye. *The crystalline (or vitreous) sphere is no longer the seat of the visual virtue (which Leonardo supposes, instead, to be situated at the extremity of the optic nerve—a place we now know, ironically enough, to be insensitive). Rather the function of the crystalline humor has been reduced to a refracting device re-inverting the array of luminous rays (or* species*) after their supposed intersection inside the pupil (which Leonardo likens to the aperture of a* camera obscura*). In Leonardo's own words:*

> *How the species are transmitted to the visual virtue with two intersections by necessity. The object* a *sends its similitude to the visual virtue through the line* ar *to part* r *of the cornea* ef*. It then enters through the pupil with its intersection at* o*, and passes to the vitreous sphere at* v *and penetrates this sphere from* v *to* q*. It then passes through the intersection* n *and terminates in* k *at the head of the optic nerve* khl *by which it is then referred to the common sense.*[1]

1 Leonardo da Vinci, *Of the Eye*, 26, in D.S. Strong (1967), p. 56.

Fig. 11 — Maurolycus' Theory of the Camera Obscura

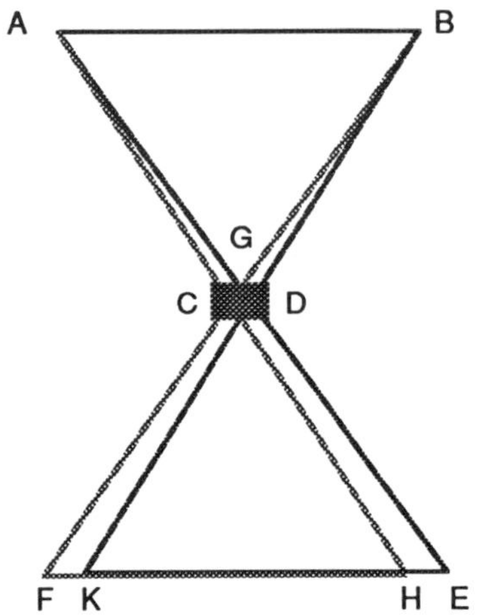

Light rays from the luminous source AB *pass through small aperture* CD *to form an inverted image* FE. FCH *and* KDE *may be considered as pyramids of radiation from vertices* C *and* D. *However, angles* FCH *and* KDE *are necessarily greater than angles* FBK *and* HAE. *Therefore, producing the rays* CF, CH *and* DK, DE *will cause the bases* FH *and* KE *(representing inverted images of the source* AB*) to increase more rapidly than the bases* FK *and* HE *(representing the separation of the two images of the source due to the width of the aperture). Consequently, as the rays are being produced, "it is...entirely possible for the spaces* FK *and* HE *to become negligible in comparison with* FH *and* KE.... *Therefore it follows that, in proportion as the rays are produced, the bases* FH *and* KE *will acquire similarity to one another and to the light-source AB, since* FH *and* KE *are figures similarly situated. And by a preceding corollary [Theorem XXI] these produced rays make it possible that both figures,* FH *and* KE, *may be thought of as one. This phenomenon will be all the more marked as the aperture* CD *becomes smaller in comparison with the source* AB. *So also in proportion as the source* AB *recedes from the aperture will* FK *and* HE *become smaller in comparison with* FH *and* KE. *Similarly we may show that the shape* FE *which is the largest, built up from the pyramid* FGE, *and the others without number built up from an infinitude of pyramids step by step come together and fuse into a form similar to the source* AB, *until they acquire a shape almost identical with it."*[1] *Thus Maurolico solved in clear and simple terms a problem that had vexed generations of perspectivists for three hundred years.*

Similar considerations hold for the other prominent perspectivists of the 16th century. Maurolico of Messina, for example, developed the first satisfactory theory of the *camera obscura. But apparently he did not regard the camera obscura as an appropriate (since purely mechanical) model of the mode of operation of the living eye.* For the latter is not sensitive to just any ray of

1 PL pp. 28-9.

light reaching it from outside but it has the special power of selective sensitivity in virtue of its quasi-intentional nature. Almost everywhere in Maurolico's important work one senses a profound theoretical ambivalence. As a result quite a few aspects of his various studies may be regarded as of pioneering quality and yet his work as a whole does not lead to a comprehensive theory of vision which in its entirety represents a radical break-through. This is because Maurolico's framework has become untenable. It confines rather than inspires. He continues to work within the traditional research program defined by the identity theory. He thus presses his novel insights into a conceptual framework in which they loose their innovating potential (just as butterflies may die in their old cocoon). For example, he develops a theory describing the refraction of light by double convex and double concave lenses. Furthermore he decides that the crystalline *is* a double convex lens. On the basis of these novel ideas he is the first scholar capable of formulating the theoretical connection between myopia and hypermetropia on the one hand and on the other the corresponding defects in the power of curvature of the crystalline lens. For the first time in history the correcting function of spectacles has been theoretically accounted for in a satisfactory way, at least in principle.[1]

Maurolico even goes so far as "to define the *pupil* [i.e. the crystalline humor] as the *lens of nature* and conversely the *glass lens* as the *pupil of art*."[2] Elsewhere he claims that "if the eye [*organum*] is a transparent body than the whole problem [of vision] is one of transparent substances".[3] But then we suddenly see him turn around and move from an apparently mechanistic to a psycho-organic Aristotelian terminology. Thus Maurolico proclaims that

> among those [ocular] structures which concern vision the aqueous or crystalline humor...occupies...the great fortress in which, as a permanent abode, the visual power resides...Here in the eye the most excellent of the humors is allotted [*sortitur*] the middle position.[4]

Thus we are back in the confines of the orthodox theoretical tradition. Apparently the mechanism of the eye cannot be studied in isolation, independently of the principal purpose it serves and to which all its parts are organically subordinated—as subjects are to their king and his God-given Kingdom—and from which, conversely, it derives its very dignity: the sublime power of

---

1 *The Photismi de Lumine of Maurolycus: A Chapter in Late Medieval Optics*, H. Crew (tr.), (New York, 1940), pp. 116-8. This work will be referred to below as PL.

2 PL p. 120.

3 PL p. 105.

4 PL p. 106.

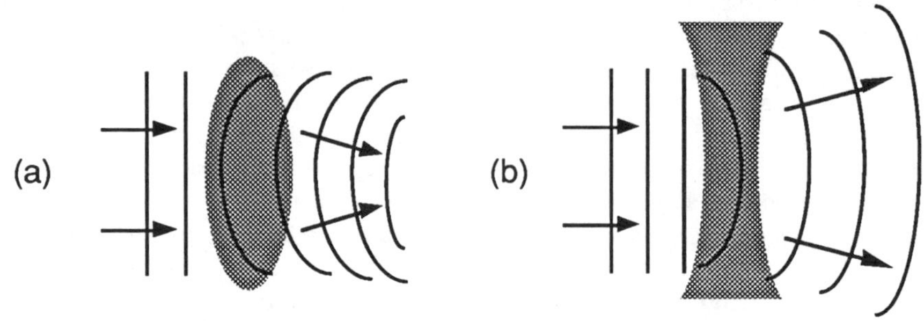

Fig. 12 — Modern view of the optical effects of converging and diverging lenses. *Wave fronts passing from air to glass are slowed down. In (a) the overall retardation is greater through the center of the lens, resulting in convergence. In (b) light waves are retarded more at the edges, resulting in divergence.*

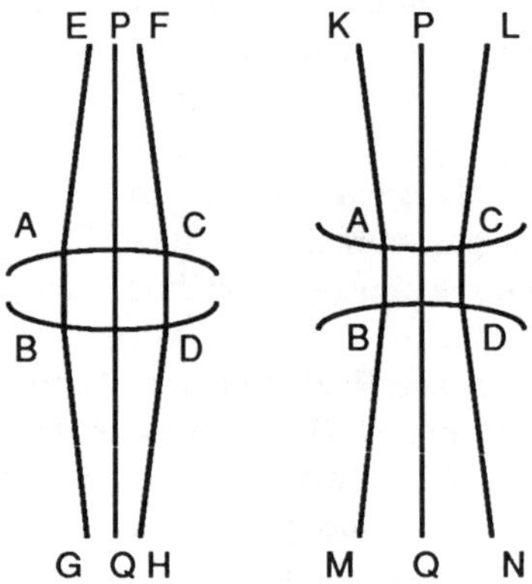

Fig. 13 — Refraction of light rays through lenses and theory of correction glasses according to Maurolycus.[1] *Double convex lenses cause rays of light to converge, while double concave lenses cause them to diverge. Thus, long-sightedness is corrected by convex spectacles, while short-sightedness requires concave lenses for its correction.*

1 PL pp. 110ff.

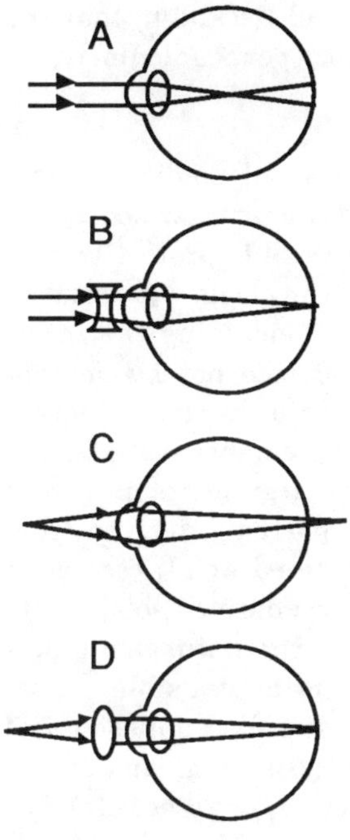

Fig. 14 — Contemporary theory of correction glasses for common defects of vision.

(A) Near-sightedness (myopia) is caused either by the eyeball being too long or the lens being too powerful (too converging). As a result the image is brought to a focus in front of the retina.

(B) This defect, common among young people, can be corrected by placing a double concave lens in front of the eye which diverges the light and makes it focus farther away from the lens.

(C) Far-sightedness (hyperopia, hypermetropia, or presbyopia) is caused either by the eyeball being too short or by a lens which is not converging enough. Consequently, the light is focused behind the retina and the actual image on the retina is blurred.

(D) This defect is common among older people when the lens has stiffened somewhat and the ciliary muscles cannot squeeze it down into greater curvature. With the aid of converging glasses the image can be properly focused.

visual perception. This purpose, and thus all its theoretical consequences, is absolutely certain. Again, it follows therefore that the theory of the mechanism of the eye should not permit the formation of inverted images. For this would clearly contradict the firm data of the phenomenology of visual perception.

This conviction is so unshakable that it can even serve in a *reductio* argument directed against certain aspects of traditional theory. Thus Maurolico confronts his new theory of the symmetry of the paths of light rays before and after passing the double convex crystalline lens with Bacon's and Peckam's old theory that the faculty of vision is sensitive only to rays perpendicularly incident upon it. But then (Maurolico argues) these rays must also emerge perpendicularly after their intersection in the center of the lens. "But", he continues, "nature is averse [*abhorruit*] to this partly because of the inconvenience of the lenticular form [for the achievement of such an end] and partly because only inverted images would be produced on account of the intersection of the rays. The idea advanced by Bacon and Peckham, that the visual rays enter the pupil [i.e., the crystalline humor] perpendicularly, is, therefore, absurd."[1]

This concludes my analysis and rational reconstruction of medieval theories of vision in terms of a persistent research program characterized by an epistemological hard core or negative heuristic and a flexible protective belt of empirical hypotheses. In an important sense the analysis has served, among other things, to expand the naturalistic thesis concerning the interplay of 'philosophical' epistemology and 'empirical' science not by demonstrating its truth but by showing its success if adopted as a research program for philosophical-historical inquiry. Thus I have supplemented a possible negative critique of analytic epistemology with *positive* arguments on behalf of naturalism, arguments which instead of being purportedly derived from evident truths or a priori insights (a fallacy to be avoided at all cost as it would undermine the very proposition thus defended) are derived solely from the illuminating force and the practical effectiveness of the naturalistic perspective as an instrument of historical analysis. It is a plausible consequence of naturalistic epistemology that the history of so-called philosophical epistemology itself should be capable of successful analysis by means of historical models specifically constructed in order to describe, or to account for, the development of *empirical* science. A positive test of this consequence should thus count as a corroboration of the theoretical proposal inherent in epistemological naturalism.

We are now in a position to pursue this line of inquiry further and to assess the impact of the scientific revolution and the mechanistic hypothesis upon the theory of knowledge, the theory of perception and the philosophy of mind. However, before doing so I should like to devote a brief chapter to discuss a controversial thesis concerning the rise of the mechanistic hypothesis itself.

---

1 PL p. 115.

V

THE MATHEMATIZATION OF PHYSICS AND THE MECHANIZATION OF THE WORLD-PICTURE GRADUALLY PREPARED IN THE DEVELOPMENT OF MEDIEVAL OPTICS RATHER THAN IN THAT OF TERRESTRIAL OR CELESTIAL MECHANICS.

As stated above the last theme of my inquiry into the history of medieval theories of perception can only be touched upon briefly. Rather than elaborate upon it I will merely summarize relevant remarks and comments scattered throughout the previous sections. At the same time this summary may serve as a convenient introduction to the subject matter to be discussed in the following chapter, viz., the impact of the mechanistic hypothesis upon Cartesian and post-Cartesian theories of vision and of perceptual knowledge.

In my view there is ample ground to suppose that the origin of (or the first intellectual moves towards) the mathematization of physics are to be found in *optics* rather than in mechanics. Alkindi's study of optics was founded upon a special philosophy of nature (or cosmology) in which optics as the science of the most fundamental natural phenomenon, viz., the universal radiation of energy, constituted the basis of all of natural science. Moreover, knowledge of this fundamental science could only be attained via the *quadrivium*, that is, via the mathematical sciences.[1]

During the Middle Ages the magico-mathematical tradition which is manifest in the above train of thought was to give the study of optics strong theoretical-geometrical as well as strong experimental impulses.[2] Both of these are defining characteristics of the later metaphysics of mechanism. It is a relatively small step from the conception that the universe is a vast network of entities radiating energy in all directions and thus exerting

---

1 Cf. Richard Lemay, *Abu Ma'shar and Latin Aristotelianism in the Twelfth Century*, (Beirut, 1962), pp. 47-8.

2 Cf. above, chap. IV, sec. 3.

influence upon each other in a way that is fundamentally capable of mathematical analysis to the mechanistic conception in which the universe of apparently qualitative changes is represented as essentially a gigantic clockwork whose laws of motion can be known with mathematical precision and which can thus be controlled and manipulated at will. But it was in *optics* (or the *theory of vision*) where for the first time this speculative program appeared to bear fruit.

To be sure, prior to Alhazen there already existed a geometrical optics. We have seen how Euclid transformed optical problems into strictly geometrical questions. But the physical interpretation of geometrical optics ran into seemingly insuperable problems. Optics in the hands of Euclid, and to a certain extent even in those of Ptolemy, appeared to be a purely abstract theoretical science as irrelevant for understanding the reality of experience as geometry itself.[1] In the controversy between the Aristotelian Avicenna and the Euclidean Alkindi the disparacy of their respective research program is still very evident. But in the process of its progressive 'physicalization' the geometrical tradition in optical theory steadily gains ground. However, a genuine break-through only succeeded after Alhazen's ingenious emendations of the respective rival research programs: for the first time the various optical research programs which thus far had developed in isolation from each other (the Aristotelian-physical-philosophical tradition, the Stoic-Galenic-medical tradition and the Euclidean-geometrical tradition) combined into a single successful program which managed to overcome the greater part of the theoretical limitations inherent in each of the respective programs taken in isolation.

However, even though Alhazen's novel theory is neither purely physical, nor purely physiological, nor purely mathematical, yet from a philosophical point of view the lasting and most influential theoretical consequence of his intellectual achievement must be that the classical argument for the theoretical irrelevance of mathematics for the understanding was muted forever. The old ideas of the magico-mathematical tradition which thus far could have been regarded as the products of an overheated imagination and as equally fanciful, speculative and unrealistic as, say, during antiquity, the Pythagorean fantasies about the *harmonia mundi*, the mathematically determined harmony of the celestial spheres, those old ideas could henceforth be used for the foundation of a *realistic science of nature*. The penetration of

---

1 Cf. chap. III, sec. 4 above. The idea that 'pure' mathematics cannot deal with, let alone be applied to, material reality is no less Platonic than it is Aristotelian. Cf. above, chap. III, sec. 2, p. 25.

this idea of mathematical optics or 'perspective' as the fundamental science in medieval Europe is clearly visible in the works of Grosseteste, of Bacon, of the artist-engineers of the 15th and 16th centuries, and of 'magicians' like John Dee or Giambattista della Porta. Thus Grosseteste, not yet influenced by Alhazen's work, expounds the doctrine of the 'multiplication of species' on the basis of received Neo-Platonic notions.[1] Moreover, Grosseteste was familiar with some of Alkindi's work, certainly with his *De Aspectibus* and possibly with his *De Radiis Stellarum* as well. Thorndike has argued[2] that the medieval physical doctrine of the propagation of light as the archetype of all natural causation is dependent upon Alkindi's cosmology of light set forth in this latter work. Thus Grosseteste's doctrine establishes a link between optics (in the sense of a general theory of light) on the one hand and a *mathematical* methodology of natural science on the other. He states quite explicitly:

> ...all causes of natural effects must be expressed by means of lines, angles, and figures, for otherwise it is impossible to grasp their explanation.[3]

All natural agents propagate their powers in all directions to surrounding bodies or recipients (*patientes*[4]) just as a luminous body radiates light in all directions like s star. And just as geometrical optics alone explains all optical effects (e.g., Euclid's rule that apparent size is proportional to the visual angle) so must geometry alone explain the varying effects of all natural transactions by analyzing the powers (or species) in terms of lines, angles and figures.

Roger Bacon, too, is quite explicit about the important connection between optics and a general mathematical physics. But Bacon had the advantage of knowing Alhazen. Thus he had witnessed the successful application of mathematical notions and techniques not only in the realm of geometrical optics proper but in the theory of vision as well. Bacon gave the first Latin account of Alhazen's theory of the eye as an optical system whose anatomical structure was characterized by precise geometrical properties. To him all of nature clearly obeyed exact mathematical laws, and knowing the laws it must also be possible to manipulate natural effects according to preconceived design. No wonder, then, that Bacon's study of optics is closely associated

---

1 Cf. Plotinus, *The Enneads* 5, 1, 6.

2 L. Thorndike, *A History of Magic and Experimental Science*, (New York, London, 1923-1956), vol. I , p. 646.

3 In *De lineis, angulis et figuris*, quoted from E. Grant (1974), p. 385.

4 Ibid. Cf. *Die philosophischen Werke des Robert Grosseteste, Bisschofs von Lincoln*, L. Baur (ed.), (Münster, 1912), p. 60.

with an original theory of *scientia experimentalis*[1] and that his enthusiastic predictions often take us into the realm of unrestrained natural magic.[2]

This latter technological aspect of the developing tradition of mathematical 'optics' (incl. the theory of vision) of course appealed to, and was in turn reinforced by, artist-engineers of the Italian urban society during the 15th and 16th centuries. To imitate nature as far as possible became the common aim of a form of art dedicated to realism and of a form of science dedicated to technology.[3] The theory of perspective was developed by artists such as Brunelleschi and Alberti. The model of the *camera obscura* was constructed and experimented with (it was also a favorite mechanical showpiece) by artists and magicians alike such as Leonardo da Vinci, Piero della Francesca, and Giambattista della Porta. Finally the properties of artificial lenses of various shapes were studied and mathematically accounted for by Maurolico da Messina who also used these artificial models for ocular dioptrics. It is clear from all this that the main ingredients of the "radical intellectual 'mutation' of which modern physical science is at once the expression and the fruit"[4] were already fully developed in the theory and practice of early medieval '*optics*'.

The above thesis becomes even more interesting if seen in the light of Cartesian philosophy. No doubt Descartes is the most radical mechanicist of modern science. In fact his mechanicism had been expressed with so much emphasis and it had been philosophically entrenched with universal claims and arguments of such apparently ineluctable force that it was almost inevitable that his original dualistic constraint would in due course be felt as an arbitrary limitation. As we will see, the materialism of the Enlightenment is an unintended intellectual product of the radical mechanicism which Descartes had declared the pivot, nay the defining characteristic, of all genuine scientific research. Descartes' earlier thoughts expressed in *Le monde* and in the *Regulae* already proclaimed this

---

1 Roger Bacon, *Opus Majus* 6; cf. L. Thorndike (1923-56), II, pp. 649ff.; and A.C. Crombie, 'The Relevance of the Middle Ages to the Scientific Movement', in *Perspectives in Medieval History*, K.F. Drew and F.S. Lear (eds.), (Chicago, 1964), pp. 45 ff.

2 Cf. e.g. *De scientia perspectiva* 3, 3, cap. 4; *Opus Majus* 5, J.H. Bridges (ed.), pp. 165-6.

3 Cf. R. Krautheimer and T. Krautheimer-Hess, *Lorenzo Ghiberti*, (Princeton, 1956), pp. 14, 229: "...imitare la natura quanto a me fosse possibile."

4 A. Koyré, 'Galileo and Plato', in *Metaphysics and Measurement*, (London, 1968), p. 16. Cf. J.H. Randall Jr., *The Making of the Modern Mind*, (Boston, 1926), pp. 220 ff., 231 ff.; A.N. Whitehead, *Science and the Modern World*, (New York, 1925).

mechanistic research program not only for cosmology but also for biology.[1] However, Cartesian science was subsequently superseded by Newtonian mechanics and the latter was to occupy a position of central paradigmatic importance during two centuries of flourishing scientific research. As a result seventeenth century mechanicism was in retrospect automatically identified with physics and astronomy. But it is good to remember that during the first half of the seventeenth century the majority of the scientific community took scarce notice of Kepler's planetary laws. As for Descartes, he plainly failed to assimilate Kepler's theoretical results to his own mechanistic program. But then, it was certainly not on *astronomy* but rather on *optics* that he had set his first hopes of establishing and applying mathematical mechanicism. The optical bias of his scientific research is clearly shown in the subtitle of his cosmological work *The World: Treatise of Light*.[2] The key to the understanding of the universe is thus to be found in *optics* rather than in celestial mechanics. Two of Descartes' early *Essays on Method*, the *Dioptrics* and the *Meteors*, deal with his optical theory as applied to the theory of vision and to the color spectrum of the rainbow respectively. His theory of vision, in its turn, is a physiological optics providing the theoretical and experimental foundation of the notorious bio-automatism hypothesis, thus introducing the mechanicist research program into the realm of biology. Finally, in the *Passions* automatism serves as the point of departure for the discussion of the interaction of body and mind. Taking all these facts into account as well as the prominence of optics and automatism in the correspondence and posthumous writings one is justified in speaking, as Caton does, of the "optical bias of Cartesian physics".[3]

This optical bias should not be interpreted as a *testimonium paupertatis*. It is not an *ad hoc* maneuver displacing a positive heuristic based upon celestial mechanics as the latter's incorporation into the mechanicist research program proved to meet with something less than astounding success. On the contrary, our historical analysis points out that the optical bias of Cartesian physics must be viewed as the progressive continuation of the old idea of optics as a privileged science which we already saw expressed in Alkindi and which was progressively reinforced by the Alhazenian synthesis and finally by the Keplerian revolution.

---

1 Cf. N. Kemp Smith, *New Studies in the Philosophy of Descartes*, (London, 1952). Also cf. T.S. Hall, 'Descartes' Physiological Method: Position, Principles, Examples', in *J. Hist. Biol.*, 3, 1 (Spring, 1970), pp. 53-79.

2 *Le monde: Traité de la lumière.*

3 H. Caton, *The Origin of Subjectivity, An Essay on Descartes*, (New Haven, 1973), p. 75.

VI

MECHANICISM AND THE RISE OF AN INFORMATION THEORY OF PERCEPTION. A NATURALISTIC RECONSTRUCTION OF (POST-) CARTESIAN EPISTEMOLOGY.

1. Keplerian dioptrics, Cartesian mechanicism, and the rise of justificationist methodologies.

The success of Kepler's dioptrics should not make us blind for the tremendous theoretical problems which the adoption of mechanicism as a comprehensive research program brought in its wake in the fields of philosophy, epistemology and theoretical psychology. It is not surprising that it was not until the days of Kepler that the dioptrical aspect of the problem of perception was (largely) solved although practically all ingredients for that solution had already been available since antiquity. For a real solution of the problem of non-radial rays which had vexed medieval scholars since Alhazen required a total commitment to the mechanistic methodology. And until Kepler's time this proved too large a step.

Even Kepler himself in several respects turns out to be no more than a transitional figure. Thus he still perceived color as a sensible *species* rather than as a mere subjective effect of the mechanical actions of extended matter as it was for Descartes and Galileo. He also suggests that the image (*species*) received by the retina passes through the continuity of the spirits to the brain, and is there delivered to the threshold of the faculty of the soul. Thus various details of his theory are recognizably traditional Alhazenian.[1] Indeed, the transition to a comprehensive mechanistic research program would require, in the first place, a radically new scientific *metaphysics* involving a fundamental critique of the traditional notion of substantial forms while replacing it with the idea that, appearances to the contrary notwithstanding, physical

---

1 Cf. J. Kepler, *Dioptrice*, prop. lxi (Augsburg, 1611), in id., *Gesammelte Werke*, W. von Dyck and M. Caspar (eds.), 18 vols., (München, 1937-...), IV, pp. 372-3.

reality is exclusively characterized by either purely geometrically determined matter-in-motion or else by intrinsically unchangeable atoms which, colorless and odorless, without heat or sound, whirl around in an infinite empty universe. Furthermore, it would necessitate an entirely new *theory of knowledge* according to which knowledge is a reconstruction, and no longer a copy, of reality. And finally, in close connection with the foregoing, a new *theory of mind* would have to be formulated according to which the mind is an information-processing device capable of autonomous organization of sensory inputs.

Philosophically all this implied a definite break with naïve (Peripatetic) realism. The privileged relation between subject and object is severed forever. The gap between being and knowing which the identity theory had attempted to bridge in a natural 'organic' way turns out to be a yawning chasm with no easy passage. The originally ontological dichotomy of appearance and (true) reality now becomes an epistemological one. As stated above, the *epistemological* dualism of appearance and Ding-an-sich is not Kantian but Cartesian in origin.[1] Finally, the domain of 'first philosophy' radically shifts from metaphysics to a theory of knowledge that is *scientifically informed*. The dominant program in the theory of knowledge now becomes the *Theory of Ideas* and *Representationism*.

Its primary task will be to formulate a theory of mind compatible with the results of the new mechanistic physics and optics and capable of adequately accounting for the discrepancies established by science between phenomenal appearances and reality as determined by intellectual or experimental means. Paradoxically enough the scientific revolution of the 17th century had thus led to the deepened insight that truth is *not* manifest. On the contrary, the origin of modern epistemology is to be found precisely in the *skeptical* implications of the supposed causal processes in perception. That is why the Pyrrhonic crisis of the 17th century went so much deeper than the traditional 'academic' skepsis of the Scholastic era. It is this new skepsis with its apparent scientific foundation and its alleged relevance to perceptual knowledge as such which Montaigne deemed insoluble and which Descartes in his *Dioptrique* and his *Principles* and Berkeley in his *New Theory of Vision* tried to defeat each in his own way.

It is no coincidence, then, that henceforth science is felt to be in need of an explicit *method* and of an epistemological *justification*. Descartes, no doubt, expresses this need most emphatically, as when he writes: "It is far better

---

1 Cf. e.g. *Principles of Philosophy*, I, LXXII, (HR I 251).

never to think of investigating truth at all, than to do so without a method."[1] But Locke, too, discusses such methodological points as the need of hypotheses (if only for heuristic purposes[2]), or the foundational significance of experiments in science[3] as well as the modest range of the knowledge derivable from them.[4] Indeed, most writers of the period including Fermat, Huygens, and of course Newton, preface their scientific works with extensive methodological justifications. Only Montaigne, the stubborn skeptic, remains utterly unimpressed with any project claiming to justify knowledge. To be sure, he also stresses the need for a criterion to tell veridical perceptions from mere appearances. But, he argues, in order to know whether the proposed criterion succeeds in its selective task we need demonstration. And for demonstration we *already* need to know what is true in our phenomenal experience. And so we are endlessly going in a circle.[5]

2. Complete demonstration in science impossible. The need of conjectural theories affirmed.

In reconstructing the intellectual history of the 17th century we have to combat two popular, and thus powerful, myths. One is the positivist myth according to which Locke (and the empiricist tradition following him) dismisses alto-

---

1 *Oeuvres de Descartes*, Charles Adam and Paul Tannery (eds.), 12 vols. (Paris, 1897-1910), X, 371, 10-16. This standard edition of Descartes' works will be referred to below as AT.

2 *An Essay concerning Human Understanding*, IV, 12, M. Cranston (ed.), (London, 1965), p. 368.

3 Ibid., p. 364.

4 Ibid., p. 365. The following passage from the *Essay* sums it up neatly: "All that I would say is, that we should not be too forwardly possessed with the opinion or expectation of knowledge, where it is not to be had, or by ways that will not attain to it: that we should not take doubtful systems for complete sciences; nor unintelligible notions for scientifical demonstrations. In the knowledge of bodies we must be content to glean what we can from particular experiments: since we cannot, from a discovery of their real essence, grasp at a time whole sheaves, and in bundles comprehend the nature and properties of whole species together." (Ibid., p. 367).

5 *The Essays of Montaigne*, John Florio (tr.), (New York, The Modern Library), p. 544.

gether a theoretical approach to science while placing his whole trust in bare phenomenal experience as the compass of human understanding and as the only true guide to the surface characteristics of natural reality beyond which the finite intellect cannot hope to reach. The other is the Popperian (and Lakatosian) myth according to which Descartes (and the Intellectualist tradition following him) is a rigid rationalist in need of conjecture nor experience but trusting solely the primitive intuitions of the mind and its clear and distinct ideas.

Such tidy dichotomies should make us suspicious of their historical adequacy. Indeed, the Cartesian myth was reinforced in modern philosophy by an excessive preoccupation with the apriorism of the *Meditations*, a pitfall which should have been avoided ever since the original interpretation Norman Kemp Smith gave in his "prentice-work" of 1902[1] and especially since his authoritative *New Studies in the Philosophy of Descartes.*[2] However, this preoccupation was not a mere hermeneutical mistake. For its deeper philosophical source is to be found in the wide-spread anti-naturalism which was one of the birthmarks of 20th century analytic philosophy. Thus philosophers only tended to know the Descartes of the *Meditations* and of other works declared to be genuinely philosophical[3] while historians of science tended to direct their exclusive attention to such works as *Le monde*, *Traité de l'homme*, *La dioptrique* and *Les météores*. The implied connections have thus been obfuscated, to the detriment of the respective exegeses.[4]

Contrary to the above myths, then, I will argue (though not at length) that for Locke the representative theory of perception (and of concept formation as well) is the foundation stone of his philosophy serving among other things as the vehicle for a defense of a theoretical approach to science in which no certain knowledge is to be obtained. On the other hand, for Descartes theoretical a priori reasoning does allow us to penetrate deeply into the hidden structures of reality and to establish certain physical principles beyond reasonable doubt. But these principles only provide a general framework for science and define a unified program of comprehensive research with an indubitable

1 N. Kemp Smith, *Studies in the Cartesian Philosophy*, (1902)

2 Id., *New Studies in the Philosophy of Descartes*, (London, 1952).

3 Thus in several editions meant for a 'philosophical' audience the *Discourse* is published without the scientific essay for which it was explicitly meant as an introduction; Descartes even called these essays "essays de cete mêthode."

4 For some stimulating contemporary exceptions to this general tendency among twentieth century philosophers cf. e.g. D.M. Clarke, *Descartes' Philosophy of Science*, (University Park, 1982); M.D. Wilson, *Descartes*, (London, 1978); N.L. Maull, 'Cartesian Optics and the Geometrization of Nature', *The Review of Metaphysics*, 32, 2 (Dec. 1978).

hard core. However, according to him, it does not allow us to derive any particular details of actual reality.[1] Thus, in order to know the 'facts of life' experience must be consulted. And so in Descartes, no less than in Locke, complete demonstration in science is impossible. Cartesian science mainly consists of intermediate theories whose relation to the apriori mechanicist principles constituting the hard core of its program is one of *compatibility* rather than of *deducibility*.[2]

Whereas in Locke we have no *experimental* access to the "radical constitution of matter" or to the "insensible parts of things" and, consequently, no access to their real essences enabling us to derive their particular properties and operations—and so, in order to close the gap between knowing and being experience must be enlightened by reason and supplemented with hypothetical reasoning; in Descartes, by contrast, we do have *intellectual* access to the real essence of material bodies and to (some of) the mechanical principles that govern their motion, but knowledge obtained by this means is too general and does not reach far enough to include the sensible (measurable) particulars—and so, in order to close the gap between knowing and being we need to supplement a priori reasoning with conjecture and enlightened experience. The two philosophical perspectives may vastly differ in their respective emphases. But the epistemological problems dictated by each one of them are very similar, indeed.

3. Ambivalence towards any alleged sources of 'immediate' knowledge. Epistemology founded on an empirical theory of the senses and the mind.

Contrary to popular interpretations, then, the philosophies of the seventeenth century are not naïvely foundationalist. But neither do they discard foundations altogether. Rather they are characterized, as we will see, by a radical *ambivalence* towards any alleged sources of immediate knowledge. And they are especially distrustful of the (suggestions of) immediate experience. Descartes dismisses it as "youthful prejudice" misleading us into uncritical acceptance of all sorts of common-sense notions that had plagued Aris-

---

1 *Discourse*, HR I 121.

2 Cf. A.I. Sabra, *Theories of Light from Descartes to Newton*, (Cambridge, 1981).

totelian science.[1] Furthermore, he shows in his *Dioptrique* to what extent visual impressions are a function of physiology rather than of the objects themselves. Ultimately the common theme underlying Descartes' epistemological expositions clearly is the notion that we neither *immediately* perceive external objects nor do we have *mediate* access to the essential attributes of surrounding bodies by means of *intentional species* flying off from them and reaching the eye.[2]

But the very same theme constitutes the fundamental point of departure of Locke's epistemology as well.[3] Because of the dualism already suggested by traditional examples[4] but inescapably dictated by the corpuscular philosophy between the sensible surface characteristics and the real essences of things neither our immediate perceptions nor, consequently, the concepts based upon them are adequate with respect to their respective objects. Our perceptions fail to be faithful copies of the real properties of things as they exist in themselves. Consequently, the abstract ideas formed to assimilate experience by combining phenomenal properties into groups are no more than arbitrary constructions of the mind which do not represent real natural kinds. Locke's attack on the Aristotelian-Scholastic identification of species with real essences is in fact an attack on naïve common-sense realism according to which sensory experience teaches us the true difference which exists between different kinds of material objects. Furthermore, there exists a close connection between Locke's rejection of naïve empiricism and his advocacy of the hypothetical method in science as well as his insistence on the merely probabilistic nature of our knowledge of physics. And here again he is in closer alliance with Descartes than is commonly appreciated. For Locke expresses his ideas by reference to "that famous clock at Strasburg" of which we only see the outward figure and motions but never the true internal constitution from which flow all those sensible ideas which we observe in it.[5] And Locke leaves us in no doubt as to the intended methodological implications of this clock metaphor. Thus, since sensible ideas are not to be identified with real essences and since there are, for the most part, no necessary connections between any two simple ideas natural philosophy is bound to be a fallible conjectural enterprise not capable of being made a demonstrative science.[6]

---

1 *Principles*, I, LXXI (HR I 249 f.); also cf. *Reply to Objections*, VI, 9 (HR II 251-3; AT 7: 437-9).

2 Cf. *Reply to Objections*, VI, 9, (HR II 251).

3 Cf. *Essay*, II, 8, Cranston (ed.), p. 82.

4 Cf. *Essay*, II, VIII, 5; II, VIII, 16-21 (Cranston, pp. 85-6).

5 *Essay*, III, VI (Cranston, p. 251).

6 *Essay*, IV, III, 10-4 (Cranston, pp. 311-3); II, XXXI, 6; IV, XII, 9-10 (Cranston, p. 365).

But it was Descartes who introduced this clock analogy, and that in order to make very similar methodological points:

> But here it may be said that although I have shown how all natural things can be formed, we have no right to conclude on this account that they were produced by these causes. For just as there may be two clocks made by the same workman, which though they indicate the time equally well and are externally in all respects similar, yet in nowise resemble one another in the composition of their wheels, so doubtless there is an infinity of different ways in which all things that we see could be formed by the great Artificer (without it being possible for the mind of man to be aware of which of these means he has chosen to employ). This I most freely admit; and I believe that I have done all that is required of me if the causes I have assigned are such that they correspond to all the phenomena manifested by nature (without inquiring whether it is by their means or by others that they are produced).[1]

Thus, whereas Locke uses the analogy to show that our *senses* do not penetrate far enough to reach the hidden nature of things, Descartes originally employed the analogy to show that *reason* does not have that power of penetration either since in many cases even a careful analysis of the phenomena allows alternative schemes of explanation compatible with the first principles of physics. Thus on Descartes' account, too, a large part of physics is shrouded in the less than certain mist of "moral certainty". But Descartes is more confident than Locke about the validity of speculative knowledge. Thus he argues:

> But they who observe how many things regarding the magnet, fire, and the fabric of the whole world, are here deduced from a very small number of principles, although they considered that I had taken up these principles at random and without good grounds, they will yet acknowledge that it could hardly happen that so much would be coherent if they were false.[2]

And he goes on to claim, even more boldly,

> that we possess even more than a moral certainty...[for]...at least the more general doctrines which I have advanced about the world and earth appear to be the only possible explanations of the phenomena they present.[3]

---

1 *Principles*, IV, CCIV, (HR I 300).

2 *Principles*, IV, CCV, (HR I 301).

3 *Principles*, IV, CCVI, (HR I 301-2).

It appears, then, that philosophers of the 17th century were in large agreement about the inadequacy and unreliability of immediate experience. Yet, their attitude was one of *ambivalence* rather than of wholesale dismissal. For experience, however discredited, must continue to provide the only true link between being and thinking. It is only the natural spontaneous suggestions of *immediate* experience that are no longer regarded as trustworthy reflections of external reality. Instead experience has to be intellectually refined, examined and evaluated. But, of course, such an evaluation has to take place in the light of some particular *theory* about the processes of perception. Thus the 17th century need to vindicate knowledge leads to a series of theories of justification which far from being a priori very often start from accepted or suggested hypotheses of a purely scientific nature. Indeed, behind Descartes' purportedly aprioristic epistemology is an *empirical* (mainly *optical*) epistemology involving an elaborate theory of the illusions of sensory qualities and allowing an accurate intellectual discrimination of what is truly objective and real in sensory information. Thus the foundation of knowledge and method is sought precisely in an elaborate theory of human consciousness which is thus scientifically *informed*.

Descartes, in particular, is acutely aware of the epistemological aporia brought about by a mathematically conceived science of nature. Galileo, too, accepted a mathematico-mechanicist ontology. But he failed to provide a quantitative theory of measurement. His account of our access to the true reality of measurable magnitudes vaguely appeals to God who wrote the Book of Nature in the language of mathematics and who made that book "stand continually open to our gaze".[1] Descartes' theory of knowledge, however, which he developed in various writings such as *Le monde*, *Traité de l'homme*, and especially *La dioptrique*, contains the first psycho-physiological account of the difference between primary and secondary qualities and, in addition, an elaborate theory of perception warranting our access to the real primary attributes of objects. The celebrated *Cogito* hardly plays any part in this founding of Cartesian science. For, surely, what is at stake here is not the question of the *ground* of mathematical knowledge but the question of its *applicability*. An a priori insight into the necessary truths of geometry is less than worthless for the establishment of a quantitative science of nature unless our senses would prove to be capable of providing reliable metric information. To be sure, according to Descartes the core of the mathematico-mechanicist program of research is irrefutably firm and indubitably true on a priori grounds. But it is a tenacious myth that Descartes would only allow a

1 Galileo, *The Assayer (Il Saggiatore)*, in *Discoveries and Opinions of Galileo*, Stillman Drake (tr.), (Doubleday, 1957), p. 276.

rigorously deductive physics to be developed in its entirety from first principles. On the contrary, Descartes repeatedly emphasizes that the gulf between the possible and the actual cannot be logically bridged as the latter domain is largely indemonstrable.[1] Only *experience* can help to bridge this gap. Now Descartes' *Dioptrique* is an *empirical* theory of justification showing that our innate geometrical notions possess objective validity and applicability in virtue of the specific mode of operation (the 'natural geometry') of our optical apparatus. To be sure, Descartes did not develop a comprehensive theory of measurement in the sense of a justification of quantitative identity. But measurement, according to him, was not something arbitrary, no pure convention. On the contrary, just as Helmholtz[2] at a later time, Descartes regarded the human body as a rudimentary measuring device: measurement begins as a 'natural' function of ordinary perceptual judgments.

4. The rise of an information theory of perception. Internal tensions of the representationist research program.

As I said, the acute need to re-establish the broken link between being and knowing led to representationist theories, which, indeed, no longer presuppose a relation of formal identity between stimulus and experience (as in Peripatetic epistemology) but which instead assume a necessary *partial resemblance* between (selected) phenomenal and stimulus properties. However, this attempt at solving the problem of perception turns out to raise more questions than it answers. And it is these questions that dictate the dynamics of the various research programs well into the 19th century. For as soon as one abandons the theory of identity the (virtually) self-evident need arises to develop an *information theory of perception*. For the sensory data are either insufficient (and so perception implies enrichment) or else partially misleading (and so perception implies selective processing and feed-back control).

But what warrant do we have for the validity of these central processes of assimilation which do not appear to be independently testable and thus are not corrigible, if necessary? And again, how shall *we* discriminate between appearance and reality when no such distinction is made in the original

1 Cf. above. Also cf. *Discourse*, VI, (HR I 120-1).

2 For Helmholtz's 'naturalism', cf. below, chs. VII-XI. Also cf. Maull (1978).

contact between object and idea? This, as I have indicated, is Montaigne's problem. And it goes a long way to explain the deep skepsis into which the 'new science' in spite of its apparent triumphs threatened to plunge the entire intellectual and spiritual world.[1]

Another apparently insuperable problem inherent in any conceivable information-theoretical account of perception was that the introduction of 'deeper' (non-peripheral) processing would quasi-intellectualize perception. Apparently the step in that direction proved too great to be taken at once. For neither the Cartesian nor the Lockean program has any room for intellectual operations at a subconscious level. In both of them seeing is pure reception while thinking is a free activity. An intermediate position is not possible. Thus the Cartesians tended to regard the processes of enrichment as automatic ones and the processes of selection as the independent work of enlightened reason leading to scientific insight. The 'empiricists', on the other hand, solved the dubious contribution of reason by seeking refuge either in idealism (Berkeley)—thus confining themselves to the absolute certainty of strictly immediate ideas—or else in naturalism (Hume)—thus confining themselves to purely natural operations without reasonable ground. As regards the process of information-enrichment: here the 'empiricists' took refuge in introspectively accessible associative mechanisms controlled by 'experience' and operating only on *conscious* contents of perceptual consciousness.

In the following I want to elaborate in more depth the internal tension of the representationist research program outlined above. This, I trust, will facilitate an understanding of the main lines of development generating the terms in which the problem of perception presented itself to nineteenth century theoreticians.

---

1 The other, historical, but by no means 'external', root of the Pyrrhonic crisis was the skeptical cultivation of a Catholic brand of skepsis inasmuch as skepsis proved to be the most effective rhetorical weapon in the hands of the contra-reformation. Cf. Luther's disarming retort (*ante datum*) :"Spiritus Sanctus non est scepticus."

5. THE REPRESENTATIONIST RESEARCH PROGRAM.

5.1. Descartes against the identity theory of perception. The necessity of an information theory of perception.

As stated above, the adoption of Kepler's dioptrics in fact spelled the deathblow for the entire research program as defined by the identity theory as such, including its weaker versions such as developed during the late Renaissance. For these 'copy theories of perception' still operated within the basic heuristic constraint that true perceptual knowledge necessarily presupposes the (formal) identity of stimulus and experience. Like their predecessors working within the traditional program, Renaissance theoreticians were thus led to regard perception as essentially nothing but the reception and absorption of true copies from the objects of perception.

However, Kepler himself was hardly aware of the momentous implication of his novel theory.[1] It was only Descartes who would first draw this radical consequence.

(1) In the first place Descartes concludes that scientific developments imply that the copy theory is untenable. That is to say, the physical impressions on the sensitive parts of the sensory organs (and finally the physiological patterns on the walls of the pineal gland) differ demonstrably both from their mechanical causes (the external objects and their various movements) and from the sensations which they in their turn generate.[2] This is not only apparent from the inversion of the optical image on the retina. But also, however perfect these pictures on the retina may be in other respects, they are no more than two-dimensional *perspective* projections of three-dimensional objects:

> These pictures normally contain ovals and diamonds when they cause us to see circles and squares.[3]

Indeed, no iconic (retinal) images faithfully 'resembling' the objects of sensation are transmitted to the brain.[4]

---

1 Cf. above p. 70, n. 1.

2 Cf. *Notes directed against a certain Programme*, HR I 443.

3 *Dioptrics, Discourse* VI, *Descartes, Philosophical Writings*, E. Anscombe and P.T. Geach (eds.), (New York, 1971), pp. 252-3; this edition will be referred to below as AG. Cf. ibid. *Discourse* V, AG 244-5.

4 *Discourse*, IV, V, AG 244.

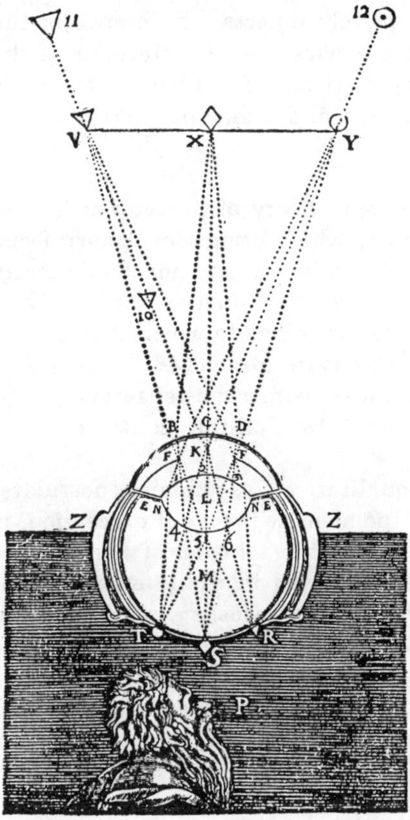

Fig. 15 — Kepler's theory of ocular dioptrics. *This theory necessitated the inversion of the optical image projected onto the retina in the back of the eye, a consequence deemed absurd by earlier generations of investigators. This illustration, taken from Descartes'* La dioptrique, *V, explains the inverted image. Rays from the sighted object* VXY *are refracted through the cornea* BCD *and the lens* L *to form an inverted image* RST *of the object on the retina. In fact the retina has here been removed for the sake of an experiment described by Descartes. It replicates Christoph Scheiner's well-known experiment of 1625. The eye, with its back coatings replaced with thin white paper or eggshell, is set in a* camera obscura. *The man looking at the eye will thus be able to actually see the inverted images cast upon the retinal screen by objects placed in front of the eye. Descartes also explains how accommodation takes place, attributing it to the shape of the lens and not to its relative position, as Kepler had theorized.*

Finally the systematic discrepancy is shown even more radically from the fact that

> the objective external realities that we designate by the words *light*, *colour*, *odour*, *flavour*, *sound*, or by the names of tactile qualities such as *heat* and *cold*, and even the so-called *substantial forms*, are not recognisably anything other than the

> powers that objects have to set our nerves in motion in various ways, according to their own varied disposition.[1]

Indeed, Descartes does not leave it at such sweeping philosophical statements. In his scientific theory of the color spectrum in the rainbow[2] he shows in concrete detail that a difference in color merely reflects a difference in the mechanical conditions of second-element particles, viz., a difference in the ratio of their rectilinear and their rotary motions.[3] The fibers of the optic nerve are sensitive to these different ratios and transmit corresponding movements to the brain.[4]

(2) Descartes goes on to argue that the copy theory of perception is also superfluous. He introduces a radical analogy which brings into sharp focus the essentially information-theoretical character of any mechanistically inspired research program into the theory of perception. Just as words

> are capable of making conceivable for us things to which they have no resemblance, why may not nature also have established a certain sign which should make us feel the sensation of light, although this sign should have nothing in itself resembling sensation?[5]

The point of the analogy is clear. The (qualitative) resemblance postulate, which was of fundamental significance in the identity theory of perception, is entirely superfluous for a scientifically acceptable solution of the psychophysiological and epistemological problem involved in the phenomenon of perceptual knowledge. In fact, far from being necessary the negative heuristic of the traditional program arises from a simple mistake.

> You must beware of assuming, as philosophers ordinarily do, that it is necessary for sensation that the soul should contemplate certain images transmitted by objects to the brain; or at any rate you must conceive the nature of these images quite

---

1 *Principles* IV, CXCVIII, AG 234; AT 8-1:322.

2 *Discourse* VIII of the *Meteors*.

3 *Meteors* VIII, in *Discourse on Method, Optics and Geometry, and Geometry*, Paul J. Olscamp (tr.), (Indianapolis, 1965), p. 337; AT 6:133.

4 Cf. *Dioptrics, Discourse* VI; cf. ibid., AG 247. In the course of his expositions Descartes clearly anticipates Müller's Principle of Specific Energies (cf. below, ch. VIII, 2).

5 *The World, Or Essay on Light*, in *Descartes: Selections*, Ralph M. Eaton (ed.), (New York, 1927), pp. 312-3 (AT 9:4). Cf. *Dioptrique, Discourse* IV, AG 243 (AT 6:12); *Principles*, IV, 197, HR I 294 (AT 8-1: 320-1).
Understandably, this analogy will make quite a career in 17th and 18th century theories of perception (e.g., in those of Berkeley and Reid) and, with new theoretical implications, it will also figure prominently in Helmholtz's cognitive theory of perception.

> differently from their way of thinking. For since they have no notion of the images except that they must be like the objects they represent, they cannot possibly explain how they can be produced by these objects, and received by the external sense-organs, and transmitted by the nerves to the brain. Their sole reason for the assumption is that they have noticed that a picture readily induces us to think of the object depicted and have thus thought we must be led to conceive of the objects that affect our senses by tiny pictures formed within our head. But we have to consider that thought may be induced by many things besides pictures—e.g. by signs and words, which in no way resemble the things signified.[1]

Indeed, Descartes pushes his argument even further. The copy theory is not only superfluous. It is even a hindrance to a scientific solution of the problem of perception, not only on the physical-physiological level (cf. above: "they cannot possibly explain, how...") but also on the psycho-physiological level. For the theory forces us to regard the physiological copy of the object as the immediate *object* of perception.[2] But this only *shifts* the psycho-physiological problem, it does not bring it closer to a scientific solution. Instead of the old (Platonic) question of how medium-sized *material things* can be the objects of a purely spiritual soul we are now faced with the question of how *physiological patterns and images* in the *sensus communis* can be the objects of a perceiving and representing mind. The gap seems smaller but it is no less deep for all that. As long as we regard the 'copy' as the object of perception we have not come one step closer to the solution of the problem of adequate perceptual knowledge than when, in a pre-scientific stage, we simply regarded the external object itself as the object of perception. For whether the mind contemplates a physiological copy or the object itself, the essential question remains: how is this object, or else this copy, capable of effecting appropriate behavior in animals and appropriate sensations in men?

---

1 *Dioptrics, Discourse* IV, AG 243.

2 For Alhazen, cf. ASP I, cap. 5, 25-6, pp. 15-6. For Witelo, cf. *Perspectiva*, in *Opticae Thesaurus Alhazeni Arabis libri septem... Item Vitellonis Thuringopoloni libri decem*, [Risner (Basel, 1572); repr. New York, 1972], III, 20. To be sure, Witelo's proposition is criticized by Kepler in his *Ad Vitellionem Paralipomena* (1604), V, 2, *Ges. Werke*, (München, 1937-59) II, pp. 151-7; [also cf. R.J. Herrnstein and E.G. Boring (eds.), *A Source Book in the History of Psychology*, (Cambridge, 1965), p. 93; this book will be referred to below as HB]. But in his *Dioptrice* of 1611, prop. LXI (*Ges. Werke* IV, 372-3), Kepler himself formulates a scarcely modified version in the tradition of Alhazen.

(3) However, this latter question does bring us closer to a solution. For if the psycho-physiological problem is translated in terms of certain mechanical cause-and-effect relations, then we suddenly realize the essential weakness of the traditional theories of perception. For they regarded the copy as the *object* and not merely as the *means* of perception. This view of the matter also prompted the demand that the immediate object of perception be an exact copy of the various qualities of the external object of perception in order that the (overall) adequacy of perceptual knowledge be explicable. But in a mechanicist research program both of these demands lose their point:
(a) The physiological 'image' (or rather: pattern of physiological activity) is not itself perceived by the mind. It only serves as a means to stimulate the mind according to certain fixed laws of human nature.
(b) Inasmuch as one no longer assumes a *perceptual* relation between the 'soul' and a physiological 'copy' in the *sensus communis*, as traditional theories had done, but only a *causal* relation between certain physiological stimuli and certain mental responses there is no longer any need to explain cognitive adequacy in terms of qualitative similarity or formal identity. Instead there arises the possibility—and in the light of the mechanicist program even the necessity—of an *information theory of perception*. In the context of this information theory the essentially unanswerable version, according to both Descartes[1] and Locke[2], of the old psycho-physiological problem is translated in terms of a scientifically more fruitful question, viz., *what physical and physiological clues determine different sensations*.[3] Or, as Descartes puts it in the *Dioptrique*:

> ...we must hold a quite similar view of the images produced on our brain; we must observe that the (only) problem is to know how they can enable the soul to have sensations of all the various qualities in the objects to which the images refer; not how they can resemble the objects. When our blind man touches bodies with his stick, they certainly transmit nothing to him; they merely set his stick in motion in different ways, according to their different qualities, and thus likewise set in motion the nerves of his hand, and the points of origin of these nerves in his brain; and this is what occasions the soul's perception of various qualities in the bodies, corresponding to the various sorts of disturbance that they produce in the brain.[4]

---

1 Cf. his *Letters to Princess Elizabeth*.

2 J. Locke, *Examination of the Opinions of P. Malebranche*, in id., *Works*, (London, 1823), IX, sec. 10, p. 217; *Essay* (1690), I, 9; IV, 3.

3 Cf. A.C. Crombie (1967), p. 74.

4 *Dioptrics, Discourse*, IV, AG 244; cf. fig. 16.

Fig. 16 — 'Natural Geometry' Cartesian style.[1] (L) *The blind man, even though he does not know the length of the sticks in his hands, nonetheless knows where the object* K *is located, because his mind, by virtue of some kind of natural geometry, can compute this value from the known distance between the two points* f *and* g *in conjunction with the known magnitude of the angles* fgh *and* gfi. (R) *Analogously, we can 'see' the distance to the object* N *as long as our mind (besides knowing a little bit of trigonometry) knows the distance* LM *between the optical axes as well as the values of the angles* LMN *and* MLN.

In sum, perceptual knowledge is not acquired through copies perceived by the soul. Rather it consists of cognitive *constructions* from entirely dissimilar physiological material which from an information-theoretical point of view, however, must be *rich* enough to allow these various constructions. On the other hand, these constructions should not be regarded as free creations of the mind. Rather, they are "ordained by Nature" to immediately follow upon, and thus be the mental correlates of, given corporeal movements in the pineal gland.

> ...we must not think that it is by means of this resemblance that the picture makes us aware of the objects—as though we had

1 R. Descartes, *Traité de l'homme*, AT XI, p. 160.

> another pair of eyes to see it, inside our brain; I have several times made this point; rather, we must hold that the movements by which the image is formed act directly on our soul *qua* united to the body, and are ordained by Nature to give it such sensations.[1]

5.2 Two radical consequences of the new theory of perception.

The replacement of the identity of perception by Cartesian mechanicism[2] also entailed the abandonment of two natural and important consequences of the traditional program. The first, *epistemological*,consequence we have already dealt with. Instead of the old idea that through the senses we acquire knowledge of the essential attributes of the world of external objects we are now faced with the notion that all we gather by this means is meaningless mechanical input which may at best contain 'cues' but which must first be decoded and interpreted. Thus naïve realism goes by the board. Instead, the new epistemological riddle created by mechanicism concerns the vexing question of how to guarantee the objective validity of our concepts within the constraints set by the mechanicist hypothesis. This question—or rather its genetic version: where do our ideas come from and how are they generated?—defines the positive heuristic of the new representationist programs in the theory of knowledge since the seventeenth century.

The second, *psychological*, consequence of the traditional identity theory involved the idea that the various distinguishable elements in the 'chain' constituting the visual process—to wit, the sensory impression, the sensation, the perception, and the conception—are not really, or formally, distinct. Rather, they should be regarded as belonging to one and the same ontological domain to be explored by one and the same field of study. However, within the mechanicist program the 'chain of perception' is no longer considered as formally homogeneous and continuous. On the contrary, there is a sudden hitch,

---

1 Ibid., *Discourse* VI, AG 246.

2 As I have indicated above, Descartes came to his mechanistic metaphysics not by way of the *Cogito* but primarily by way of his interest in (physiological) optics, and, in general, in the question of how to account in purely mechanistic terms for the relation between perceiver and perceived within the context of a Keplerian dioptrics. Cf. N. Kemp Smith, *New Studies in the Philosophy of Descartes*, (London, 1952), pp. 21 ff.

or a qualitative break, between the physiological impression on the one hand (i.e., either the peripheral or the central response) and on the other the sensation as an element of consciousness. This ontological 'leap' provokes and defines the psycho-physiological problematic which has dominated Western thought up to the present day. Indeed, the history of psychology can be described, in part, as a process of progressive differentiation of the various elements of the perceptual chain. In Reid the sharp distinction follows (already prepared in Malebranche and Hartley) between *sensation* (as a purely immanent mental event, comparable to pain) and *perception* (as something already characterized by conceptual definition as well as involving objective reference). Again, at a later stage *concept formation*, or conceptual apprehension, as involving a conscious cognitive activity is sharply distinguished from the unconscious processes underlying the formation of developed perceptions.

5.3 The negative heuristic of the Cartesian research program. Dualism of thought and sense. Descartes' information theory not a cognitive theory of perception.

So far in the discussion of Descartes we have noted that the introduction of the mechanistic hypothesis in physiological optics seems to necessitate a full-blown information-theoretical conception of the mind in order to bridge the gap, now exposed, between the mechanical input of a strictly neuro-physiological kind on the one hand and certain mental responses in the form of adequate sensations on the other. However, as we shall see, the intellectual leap to such a novel and, indeed, revolutionary conception of the human mind proved much too great to be taken at once. At any rate, for the time being this conception is blocked by the negative heuristic of the mechanicist program itself, viz., the *Cartesian dualism of sensing and thinking, of seeing and judging.*[1] Put another way, we could say that what later on in the history of

1 Indeed, one might speculate that new research programs tend to be generated by the internal tensions built into the old preceding program. Thus the mechanicist program was the fruit of the internal tension characteristic of the identity theory program since the synthesis achieved by Alhazen. In its turn, mechanicism as the dominant metaphysics leading to representationism in the theory of perceptual knowledge similarly created an internal tension inasmuch as it dictated on the one hand a dualism between judgment and sensation and on the other an information theory of perception suggesting

psychology used to be distinguished as sensation and perception Descartes and his followers still regarded as indistinguishably identical.[1]

According to Descartes the mind is pure and ever-active thought or consciousness. The body, by contrast, is no more than extended matter-in-motion. There is nothing 'in between'. The interaction between the two substances is thus immediate, there being no vehicle for interaction that could bridge the gap as it were. Consequently, there is no room for a *genuinely* information-theoretical account of post-retinal or post-terminal[2] interpretive processes of a subconscious character.

In the following I will examine how the Cartesian doctrine of the absolute distinction between thought and sense dominated, and *constrained*, the post-Cartesian development of the theory of perception. Accordingly, perception had to be interpreted in terms of purely passive sensations without the intervention of 'ratiocination' or of any quasi-intellectual processes. To what profound extent the history of perception after Descartes has struggled with this conceptual constraint built into the mechanico-representationist research program and in fact *blocking the rise of a genuinely autonomous psychology of perception* most clearly transpires from the interesting theory of perception which was developed by Malebranche (1638-1715). A detailed discussion of his views may thus serve as a representative test case of my rational reconstruction of the relevant historical episode.

---

an interpretive view of phenomenal experience. This theoretical contradiction ultimately led to the recognition that active cognitive strategies are inevitably operative in the subconscious processes leading to perception. This, in short, will be the gist of the Helmholtzian program.

1 Sensation might provisionally be defined as the product of purely passive psychic receptivity of which we are largely unconscious. Thus it takes the training of a perspective designer or painter to become aware of the purely two-dimensional information concerning the visible world which reaches us through our retinas. By contrast, perception is a product of active mental (we might even say: quasi-intellectual) processes operative either by virtue of our "natural constitution" (as in Reid) or else involving cognitive strategies that are essentially adaptive and guided by experience (as in Helmholtz). These information-handling post-retinal processes take place largely subconsciously. Consequently, their conscious products are *phenomenologically* indistinguishable from the 'passive' sensations constituting the so-called pure data of sense.

2 In Descartes' model the pineal gland is the terminal point of afferent physiological activity.

6. MALEBRANCHE AND THE CARTESIAN RESEARCH PROGRAM INTO OPTICAL EPISTEMOLOGY.

6.1. Ambiguities in Descartes' theory of sensory judgment. Lack of a genuine (cognitive) theory of information processing.

During the 17th and the 18th centuries the psycho-physiological problem of the processing of sensory information was principally focused upon the problem concerning the perception of distance and magnitude. As a result of ongoing research and theoretical activity particularly in this area the distinction between seeing and judging was developed and sharpened considerably in due course.

Descartes mentions four 'means' or cues of visual distance discrimination.[1] They are: (1) accommodation, or information concerning the "shape of the eye"[2]; (2) convergence; (3) clearness and luminosity; (4) collateral information concerning actual dimensions. But Descartes' theory is not so much a theory of information *processing* in the proper sense but at best a psycho-physiological *correlation* theory without a detailed insight into the nature of the information-handling processes. Clearly, Descartes is very reluctant to allow any genuinely mental activity regarding the transformation from pineal patterns to conscious sensations. All he is willing to admit are immediate and naturally determined transactions between body and mind. In this respect, then, he even comes fairly close to Hume's concept of natural belief.

> ...it appeared to me that I had learned from nature all the other judgments which I had formed regarding the objects of my senses, since I remarked that these judgments were formed in me before I had the leisure to weigh and consider any reason which might oblige me to make them.[3]

Descartes' theory of judgment, however, is steeped in considerable ambiguity and even inconsistency. For on the one hand he envisages appearances (or ideas as immediately apprehended) as in no way tinged by the influence of judgment so that appearances are infallibly true *considered in themselves*:

---

1 *Dioptrics, Discourse* VI, AG 249-52.

2 What is meant is the shape of the lens, of course.

3 *Meditations* VI, HR I 188-9; cf. N.K. Smith (1952), p. 251.

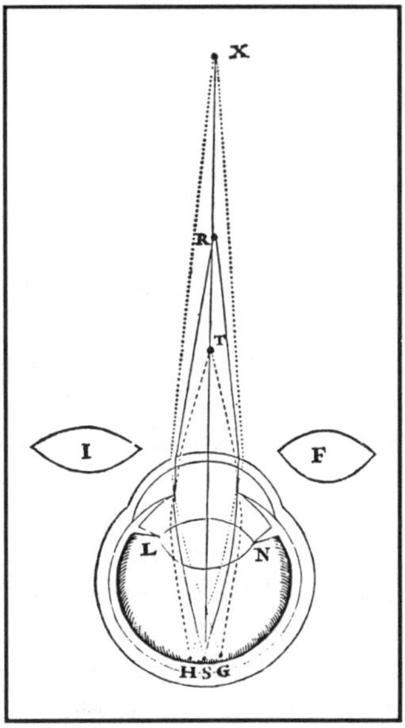

Fig. 17 — Descartes' theory of accommodation. *While the point* R *is brought to a sharp focus and projected onto the retinal point* S, *points* X *and* T *will remain completely out of focus. Thus the ray* TL *will go to* H, *while the ray* TN *will go to* G; *alternatively, ray* XL *will go to* G, *while ray* XN *goes to* H. *In order for the more distant point* X *to be focused onto the retina, the crystalline humor* NL *has to assume a flatter shape (as in I); conversely, in order for the closer point* T *to be "distinctly represented", the crystalline has to assume a more curved shape (as in F).*
(From Descartes, Tractatus de homine, *AT XI).*

> If ideas are considered in themselves, and not as referred to some other thing, they cannot, strictly speaking, be false. For whether I imagine a goat or a chimera, that I am imaging the latter is no less true than that I am imaging the former.... There thus remain only our judgments: and it is in respect of them that I must take diligent heed lest I be deceived.[1]

Thus, in the realm of appearance what seems, is. And it is only in judgment that we may go awry. For judgment involves an *aliquid amplius* (or an amplification) taking us beyond the realm of pure appearances, or "clear perceptions".[2] However, one of the main themes of the *Dioptrics* is precisely that judgment also determines what objects (their size and distance) will *appear* to us to be.

> That magnitude, distance and figure can be perceived by reasoning alone, which deduces them one from another, I have proved in the *Dioptrics*.[3]

1 *Entretien avec Burman*, Adam's edition, p. 24; AT V, p. 152.
2 Cf. *Principles*, I, 45-6, HR I 237.
3 *Reply to Objections*, VI, HR II 252.

Thus, it would appear, the 'pure' data of sense are also far from virginal and, consequently, they are liable to vice, the vice of error in this case. But this would seem to contradict Descartes' theory of sensory illusion as something never residing in phenomenal facts alone.
Furthermore, if "seeing distance" involves the use of "natural geometry", how is it that lower animals incapable of the art of reasoning can "see" objects at a distance?
Again, if the "reasoning" allegedly involved in perception is incorrigible and belongs to the province of natural belief rather than to that of pure thought, it would seem altogether misleading to speak of "judgment", of "natural geometry...like that made by surveyors"[1] and of a "grade of sensation...clearly depend[ing] upon the understanding alone."[2]

These terminological confusions, however, result from Descartes' attempt to tackle the momentous problem—indeed, he was the first one to do so in an informed way—of how we can come to know the independently real world *indirectly*, that is, by means of appearances that are distorted relative to the cerebral patterns from which they have been derived. As a result of the union of body and mind our immediate apprehensions of sensory information not only differ from the peripheral images in the organs of sense but also from the cerebral images in the pineal gland, the seat of the *sensus communis* where all available information concerning the external world comes together. And yet sense experience is essential to our knowledge of the world. An adequate theory of knowledge, therefore, must account for the *representative* character of the data of our immediate awareness. And this will require a detailed theory of the processing of sensory information. Descartes only laid the groundwork for such a theory. As we will see, it was left to Malebranche to develop the theory considerably and to disentangle the various confusions we have noted above.

That Descartes only provided a theory of information correlation, not a theory of information processing, is clear from the following.
(1) In contrast to later theoreticians Descartes in the *Dioptrics* speaks of "seeing" distance rather than "judging" distance in the case of all cues mentioned above except the last one. Thus, Descartes regarded as immediately-given-in-visual-perception all sorts of conscious data or sensations without discriminating between the data derived from purely retinal information (e.g., clearness and luminosity of the array of light received upon the retina) and data derived from additional non-retinal information (e.g.,

---

1 *Dioptrics, Discourse*, VI, AG 250.
2 *Reply to objection*, VI, HR II 252.

convergence and accommodation). Later theorists would distinguish between retinal and 'deeper' non-retinal information-handling processes. Also, they would confine 'seeing' in the proper sense, i.e., the given in visual perception, to what is contained in, or derivable from, the retinal image. Obviously, this would result in greater attention being given to those 'deeper' coordinating and learning processes, e.g. to the question how tactual and kinæsthetic information is coordinated with visual information. The empiricist learning theories and theories of concept formation (Berkeley's in particular) would be especially concerned with processes of that kind.

(2) Descartes' primary interest being in information *correlation* also transpires from his failure to discriminate between conscious (e.g., retinal definition) and unconscious (e.g., accommodation) inputs and, consequently, from his failure to formulate respectively different learning or processing theories. Berkeley, by contrast, would later restrict the possible relevance of any postulated learning processes to conscious inputs alone.

> It is evident [he said] that no *idea* which is not itself perceived can be...the means of perceiving any other *idea*.[1]

While the cogency of Berkeley's argument was merely apparent and although his doctrine would hinder rather than advance the development of psychology, it used to be regarded as an incontrovertible principle from Porterfield down to Titchener and Schlick.

6.2 Malebranche's theory of visual distance discrimination and of apparent magnitude.

Now Malebranche, in sharp contrast to Descartes, appears to be acutely aware of the intolerable lack of a detailed information *processing* theory in the Cartesian theory of perception. He offers approximately the same list of means ("moyens"), or cues, of visual distance discrimination as Descartes had given. He explicitly adds a further cue, viz., interposition, or the number and kind of intervening objects. But Malebranche does not call visual distance discriminations based upon these 'means' the result of 'seeing' but of 'judgment'. However, while he sometimes uses the expression "natural

1 George Berkeley, *A New Theory of Vision, and Other Writings*, with an introduction by A.D. Lindsay, (E.L. 483, London, 1957), X, p. 15.

judgment", he also speaks of "compound sensations".[1] This terminological ambiguity, which we will shortly examine, will prove to be highly significant.

However, deferring for the moment the problem of the exact nature of the judgments allegedly involved in visual distance discrimination we must first explore Malebranche's even more interesting observations concerning the visual perception of magnitude. Even *apparent* magnitude, he claims, is estimated or inferred. Thus, in connection with the cue of interposition, listed as the sixth means of visual distance discrimination[2], he asserts that a tower, if seen beyond intervening fields and houses, appears larger than if seen in isolation even though the size of the retinal image is the same in both cases. In general, he concludes,

> We judge the magnitude of objects by how far we believe them to be removed from us. And the bodies we see between ourselves and the objects very much assist our imagination in judging their distance from us.[3]

He then goes on to derive from this an information-theoretical explanation of the varying apparent magnitude of the moon at the horizon as compared to its apparent size when in midheaven. When the moon rises we "judge" it to be very distant:

> We judge it to be much more removed from us when it rises than when it is very high above our horizon.[4] [In fact] it...even appears to us to be beyond the perceivable horizon or beyond the lands that bound our visual field.[5]

While when the moon is in midheaven and the cue of interposition virtually fails, we "judge" it to be much closer:

---

1 "Sensation composée". Cf. *Recherche de la Vérité*, Liv. I, Chap. VII, 4, *Oeuvres Complètes de Malebranche*, publié par MM. de Genoude et de Lourdoueix, Tome premier, (Paris, 1837), p. 17. This edition of Malebranche's book will be referred to below as RV.

2 RV IX, III, p. 22.

3 "Nous jugeons de la grandeur des objets par l'éloignement où nous les croyons; et les corps que nous voyons entre nous et les objets aident beaucoup notre imagination à juger de leur éloignement." (Ibid).

4 "[N]ous la jugeons beaucoup plus éloignée de nous lorsqu'elle se lève que lorsqu'elle est fort haute sur notre horizon." (RV, p. 23).

5 "[E]lle nous paraît...même au delà de l'horizon sensible ou des terres qui terminent notre vue. " (RV, pp. 22-3).

> we judge it to be only about half a mile away from us: or seven or eight times as high as our houses.[1]

Now as a consequence of these differential *judgments* of visual distance, Malebranche assures us, the horizon moon "appears" to be much larger: "the moon appears to us to be much larger when [it rises]."[2]

6.3 Regis contra Malebranche's information theory of perception. 'Corroborated empirical excess content' of the Cartesian program according to Malebranche.

No doubt this explanation of the moon illusion according to which a mental estimate of distance should affect the actual *perception* of magnitude is both shocking from an epistemological point of view and revolutionary from an information-theoretical point of view. In fact, Malebranche acknowledged as much:

> One immediately expects that only a very few people will be found who are not shocked by this general proposition here advanced, viz.: that we have not a single sensation of external objects that does not contain one or more false judgments. We know very well that most people do not even believe that there is any judgment true or false to be found in our sensations.[3]

Indeed, Regis objected in his *Système de Philosophie*[4] that an astronomer's knowledge of the moon's actual distance and magnitude by no means alters his perception of its apparent magnitude. Consequently (he argues) the

---

1 "[N]ous ne la jugeons qu'environ à une demi-lieue de nous: ou sept ou huit fois plus élevée que nos maisons." (RV, p. 23).

2 "[L]a lune nous paraît plus grande lorsqu'elle [se lève]." (RV, p. 22). The text actually reads: "...when it is very high above the horizon" (... lorsqu'elle est fort haute sur l'horizon"). But this is clearly an erratum which indeed (as Cowper, impressed with the misleading potential of a single erratum, once put it) almost "knocks out the brain from the whole passage".

3 "On prévoit bien d'abord qu'il se trouvera très-peu de personnes qui ne soient choquées de cette proposition générale que l'on avance, savoir: que nous n'avons aucune sensation des objets de dehors qui ne renferme un ou plusieurs faux jugements. On sait bien que la plupart ne croient pas même qu'il se trouve aucun jugement ou vrai ou faux dans nos sensations." (RV, I, XIV, I, p. 33).

4 P.S. Regis, *Système de Philosophie*, 3 vols (Paris, 1690).

psychological cause adduced by Malebranche is ineffective. Instead one should trace the moon illusion to the varying size of the retinal image projected by the moon in its various positions due to differential refraction.

This theory was not original with Regis. Malebranche already observed in the *Recherche* that this was the common view shared by "a very large number of philosophers"[1]. In fact, in so far as the theory attempts to refer the entire content of visual perception back to the retina it shows unmistakable vestiges of the traditional research program in the theory of perception with its emphasis on the eye as the seat of vision and on the content of vision as being similar to the *species* received in the eye. However, in the same passage cited above Malebranche had already effectively refuted the opposing view by arguing that refraction does not affect the size of the retinal image but only the apparent elevation of celestial bodies above the horizon. Furthermore, he added, astronomical measurements of the angular diameter of celestial bodies show that the moon's diameter *increases* in proportion to the moon's removal from the horizon and so, paradoxically, we see the moon as smaller while its retinal image *grows* in size.[2]

But in his *Réponse à M. Regis* Malebranche also seizes the occasion for a positive defense of his own theory. He especially elaborates, and even sharpens, his view that *all* visual sensations are effected by judgments of distance. The apparent magnitude even of objects in our immediate environment requires the operation of complex post-retinal information-handling processes. A dwarf, two paces from us, certainly appears much smaller than a giant, three times as large, who is at six paces from us, *even though the retinal images projected by them are equal in size*. Since, therefore, the inequality in the apparent magnitudes cannot be traced to the size of the retinal images, it must arise from the perceived inequality in the apparent distances.[3]

Malebranche goes on to show that in contrast to the opponent's research strategy the Cartesian research program can claim what would, in modern terms, certainly qualify as 'corroborated empirical excess content'.[4] He describes

---

1 "[U]n très-grand nombre de philosophes" (RV, p. 23).

2 Ibid.

3 *Réponse à M. Regis*, I, 3; in *Oeuvres de Malebranche*, P. Costabel, A. Cuvillier and A. Robinet (eds.), vol. XVII-1, (Paris, 1960), pp. 264-5. I will refer to this edition as OM.

4 Cf. I. Lakatos, 'Falsification and the Methodology of Scientific Research Programmes', in id., *The Methodology of Scientific Research Programmes, Philosophical Papers*, Vol. I, J. Worrall & G. Currie (eds.) (Cambridge, 1978) pp. 8-101.

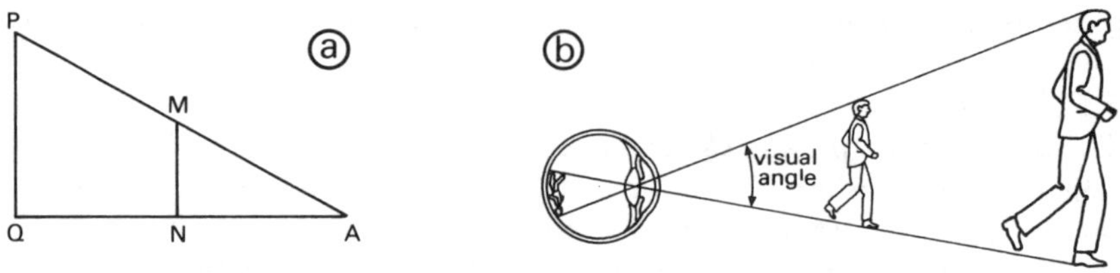

***Fig. 18 (a) — Diagram by Malebranche** demonstrating his view that vision is thoroughly affected by judgments concerning the distance of visible objects and that apparent magnitude even of objects in our immediate environment requires the operation of complex post-retinal information-handling processes. Thus although the object* PQ *is double the size of the object* MN *and twice as distant from the eye at* A, *yet since it is seen under the same angle, it must trace on the retina an image equal in size to that traced by* MN. *Nevertheless, when we notice the distance the apparent magnitude of* PQ *is about double that of the object* MN. *Consequently, the inequality in the apparent magnitudes cannot be traced to the size of the retinal images. It must therefore arise instead from the perceived inequality in the apparent distances.*[1]

***Fig. 18 (b)** — The same principle shown in a more concrete fashion. Two objects with the same visual angle have the same retinal size.*

an experiment which optically isolates luminous objects in the field of view. When a glass blackened with soot is thus used to observe the horizon sun or moon they appear to be no different in size than when viewed in the meridian.

> If the sun is at the horizon, the interposition of the glass will make it appear about two times as close and four times as small or thereabouts: for precision is not necessary here. But when it is very high above the horizon, the glass will not produce any significant change neither in its distance nor in its apparent magnitude.[2]

---

1 OM pp. 264-5.

2 "Si le Soleil est dans l'Horison, l'interposition du verre le fera paroître environ deux fois plus proche, et quatre fois plus petit ou environ: car ici la précision n'est pas nécessaire. Mais s'il est fort élevé sur l'Horison, le verre ne produira aucun changement considerable ni dans sa distance, ni dans sa grandeur apparente." (Ibid., I, 6; OM XVII-1, p. 266).

These novel effects are precisely anticipated by the Cartesian information-theoretical account as developed by Malebranche. They thus help to qualify that research program as a 'progressive' one.[1]

6.4 Tensions between the positive and the negative heuristic of the Cartesian research program. The negative heuristic at work in Malebranche's theorizing.

Even more important in the present context, however, is Malebranche's reply to Regis' objection that an astronomer's knowledge of the moon's actual distance does not affect his apprehension of apparent magnitude. For this reply touches upon the question, essential to our present discussion, as to the *nature* of the judgments allegedly involved in ordinary perception. As against Regis Malebranche emphasizes the important point that in order to exercise any influence upon our perceptions of magnitude *distance must be actually perceived by, or at least impressed upon*[2], *the senses*. Thus he writes:

> ...a distance which is not actually perceived by the senses must be counted as nothing, or it cannot serve as the foundation for a natural judgment which arises in us concerning the magnitude of objects.[3]

Now the reason for this is the absolute distinction between sensible as opposed to merely intellectual knowledge with respect to their origin (or their psychological aspect) as well as their validity (or their epistemological aspect). For sensible effects are produced in virtue of the laws of the union of body and

---

1 Cf. I. Lakatos, (1978).

2 This qualification is in the spirit of Malebranche's overall theory since he allows that visual distance discrimination utilizes cues which we do not consciously perceive (e.g., the data involved in accommodation). Thus the meaning of the quote from Malebranche cited above cf. following note) is not the same as that of Berkeley's "evident" principle that no idea not itself perceived can be the means of perceiving any other idea. On the contrary, Malebranche does not mean to draw a distinction here between conscious and unconscious material but rather between intellectual and sensible material. And the latter is defined not in terms of what is available *in consciousness* (to be determined by introspection), but rather in terms of what is *physiologically present to the mind* (to be determined by studying physiological activity).

3 "...une distance, qui n'est point actuellement aperçûë par les sens, doit être contée pour nulle, ou ne peut servir de fondement au jugement naturel qui se forme en nous de la grandeur des objets." (ibid., I, 4; OM XVII-1, p. 265).

mind. But intellectual knowledge is gained through free activity of the mind alone.

Clearly, underlying this distinction drawn by Malebranche is the Cartesian dualism of sense and thought. And this epistemological dualism, in its turn, we have recognized as nothing but the twin version of the ontological dualism of body and mind. As an offshoot of Cartesian mechanicism this epistemological dualism barred Descartes, as we have pointed out, from developing an adequate theory of genuine information *processing*. Instead, he was constrained to formulate no more than an unsophisticated and unsatisfactory theory of information *correlation*. To invoke subconscious and even quasi-intellectual operations transforming internal representations in accordance with complex optical principles was simply forbidden by the negative heuristic of Cartesian mechanicism inasmuch as the latter implied a radical dualism of body and mind. And yet at the same time an information-theoretical solution precisely along these lines seemed to be strongly suggested by the positive heuristic of that very same program.

Thus Descartes, in elaborating upon 'natural geometry' Keplerian style, apparently did assume subconscious computation determining visual distance by utilizing the cue of convergence of the optical axes. Yet he had no easy way of accounting for the processes involved. He finally attributed them to an "act of thought"[1] which he described—inconsistently, it seems, and certainly in conflict with his later teaching in the *Meditations* and the *Principles* of the mind as essentially active and conscious of all its activities—as "implicitly [containing] a reckoning like that made by surveyors"[2] and yet involving no more than "a *simple* act of imagination".[3] If this is already obscure in itself, it is not at all clear how (as Descartes goes on to explain) natural geometry "by means of a single eye if we make it change its place"[4], i.e. by means of *sequential* rather than simultaneous impressions, could possibly constitute a simple act of imagination.

Seen in this light the terminological novelty introduced by Malebranche is highly significant. For whereas Descartes dealt with natural geometry as constituting a "*simple*" act of the imagination, Malebranche speaks of natural judgments being "*compound*". This terminological change reflects, I believe, a progressive shift in the positive heuristic of the developing

1 "Action de la pensée"*Dioptrics*, *Discourse*, VI, AG p. 250.

2 Ibid.

3 Ibid.; emphasis added.

4 Ibid.

representationist research program towards a more detailed *psychological* theory of post-terminal information-handling processes.

As a result, the internal tension built into the mechanico-representationist research program is much more conspicuous in Malebranche than it was in Descartes. For on the one hand Malebranche was keenly aware of the hiatus in Descartes' theory of vision and of the need to attribute to the mind (or to *some* mind, at any rate) an organizing role with respect to the data of sense. While on the other hand he could not free himself from the dualist doctrines defending the Cartesian conception of science. Consequently, all quasi-intellectual processes, which he himself had just identified and analyzed with great care, are projected back into the sphere of the purely sensible. Thus all *perceptions*, although apparently intellectually organized and being at any rate very complex, are yet to be regarded, after all, as no more than mere passive *sensations*, 'compound' sensations, to be sure, but nonetheless passively produced.

6.5 Rational reconstruction of Malebranche's occasionalism. Divine intervention and the computer analogy.

The impossible tension the ambiguity discussed above naturally generated in his theory Malebranche has sought to resolve by a *deus-ex-machina* theory which in the history of psychology met with so much scorn and derision that it even eclipsed the extraordinarily advanced character of Malebranche's theory of perception. However, the history of ideas deserves a less narrow-minded historiography. Malebranche's universally denounced so-called 'occasionalism' should not be regarded as a scientific aberration. Nor should it be laid to rest in the bric-à-brac collection of theological absurdities sprung from the heated imagination of a deranged priest who had lost his intellectual balance since the scientific revolution. On the contrary, it should, and it can, be reconstructed as a *rational* development of a research program being confronted with its own conceptual limitations in that it provokes substantial questions (concerning the interaction of body and mind) which in principle it cannot answer *within* the mechanistic framework of science as conceived by it. In other words, Malebranche's occasionalism should be regarded as a first and serious, if obsolete, attempt to provide an adequate solution for the exceedingly complex information-theoretical problem of organizing 'ratiomorphic'[1] activities apparently operative in the production of even primitive perceptual knowledge. This problem was the immediate result of the mechanistic strategy being applied to the field of perception. And it led to a thorough theoretical aporia at two related philosophical frontiers.

First, the *psycho-physiological* frontier concerning the relation of two essentially different entities, viz., body and mind: how is it possible at all for any transactions to occur between levels of reality so utterly different? Moreover, who, or what, controls these apparent interactions? For, surely, these interactions prove to be so incredibly complex that a simple correlation theory as in Descartes clearly does not suffice to explain these psycho-physiological events. On the other hand, the more we learn to appreciate the literally mind-boggling complexity of the information-handling processes involved, the more implausible becomes a purely intellectual solution of the mind-body problem. This virtually unsolvable theoretical dilemma finally drives Malebranche to embrace the doctrine embodied in his occasionalism: God himself takes care of the apparent interactions of a seemingly intellectual character between body and mind. Hence Malebranche's dictum that ultimately we "see all things in God."

From a modern point of view this looks like a pretty bizarre theory. But we ought to keep in mind, of course, that the god of the 17th century philosophers belonged to the natural furniture of the universe. As a matter of fact, Malebranche's contemporaries did not take exception to his theory on *ontological* grounds. Thus John Locke, in his acute critique of Malebranche's theory[2] declared the "vision en Dieu" to be unscientific not because it was based on theological presuppositions but rather because, according to him, it manifestly failed to account for the adequacy of perceptual knowledge, i.e., because this theory failed from an *epistemological* point of view.

Secondly, however, and much more importantly, we should heed the fact that the terminology used by Malebranche in his description of the complex information processing he had analyzed is not so much inspired by theological concerns but rather constitutes a striking anticipation of the

---

1 Cf. E. Brunswik, 'Ratiomorphic Models of Perception and Thinking', *Proc. 14th Internat. Congress Psychol.*, M. Mailloux (ed.), (Montreal, 1954); also id., 'Scope and Aspects of the Cognitive Problem', in *Contemporary Approaches to Cognition, The Colorado Symposium*, (Harvard, 1957).

2 J. Locke, *Examination of the Opinions of P. Malebranche*, in id., *Works*, (London, 1823).

modern *computer analogy*. And this analogy, as we know, has proved to be extraordinarily fruitful in the history of cognitive psychology.[1]

The second aporia brought about by the application of the mechanicist research program to the field of perception concerned, as we have noted, the *epistemological* problem of the objective basis of human knowledge. In Malebranche's remarkable theory of sensory judgments[2] this problem comes about in the following sequence of steps leading to skepticism with regard to the senses.
(1) The sensory judgments are pervasive and enter into virtually every perceptual cognition.
(2) Although without the intervention of natural judgments no 'veridical' perception (in the strictly pragmatic sense of allowing adaptive conduct in complex physical environments[3]) is possible[4] yet these judgments are intrinsically fallible and a special source of illusion[5] because the data on which they depend are valid only relative to the body and its poor sensory apparatus.
(3) Finally, these judgments are involuntary and irresistible. The soul plays no active role in their formation.[6] In fact they represent deep post-terminal levels of information processing which are as 'natural' and God-given[7] as the occurrence of specific sensations (of secondary qualities) given certain physical impressions upon our sense organs.

> That is why I have called those kinds of judgments *natural* in order to indicate the fact that they arise in us, without us, and even in spite of us.[8]

Consequently, Malebranche's theory of vision serves to buttress his favorite methodological and epistemological view that experience and science have no business with each other. Whoever identifies sense and science misrepresents both. Whereas science is concerned with the discovery of invariant truths about reality, sensory experience only leads to practical knowledge necessary for the preservation of life. And whereas scientific

---

1 Leibniz, himself a computer scientist of the first hour, further developed the computer analogy and, moreover, related it to a revolutionary theory of the subconscious.

2 "Jugemens des sens" (RV I, VII, p. 17).

3 Ibid., p. 24.

4 Ibid., p. 17. Similarly Berkeley, cf. A.A. Luce, *Berkeley and Malebranche, A Study in the Origins of Berkeley's Thought*, (Oxford, 1967), pp. 36, 44.

5 RV IX.

6 Ibid., n.1, p. 20.

7 Ibid., p. 24.

8 "C'est pour tout cela que j'ai appelé *naturels* ces sortes de jugements, pour marquer qu'ils se font en nous, sans nous, et même malgré nous." (ibid.).

knowledge is *consciously* produced by the *individual* mind, perceptual knowledge comes about *unconsciously* through the intervention of *supra-individual* operations.

Thus, even though Malebranche seems to realize more acutely than Descartes that visual perception in general involves more than mere seeing, he ultimately refuses, just like Descartes, to develop a realistic theory of information-handling processes capable of utilizing subconscious retinal and non-retinal cues for the development of complex but adaptive perceptual object hypotheses. For the 'natural judgments' he invokes are the very opposite of finely tuned cognitive strategies designed to minimize error and to maximize perceptual reliability. Rather, they occur automatically, as instinctive and instantaneous reflexes, according to fixed and immutable laws governing the union of body and mind. They apparently involve "instantaneous reasonings"[1] which vary with every movement of our eyes, but in truth they are operations "all of which are created by an eternal act".[2] Their surprisingly fortunate results can be explained only by reference to omnipotent Providence, not to mortal intelligence:

> One will feel the hand of the Almighty and the unfathomable depths of his wisdom in his Providence.[3]

Indeed, both the purposefulness and the exceeding complexity of these activities defy even an explanation in terms of merely natural processes. Instead they call for a non-mechanical agency, an Aristotelian entelechy, nay, the omnipotent God himself:

> I say God and not nature; for that vague term *nature*, which is so fashionable these days, is no longer suitable to clearly express what we mean by Aristotle's *entelechy*.[4]

However, since God always acts in accordance with the same laws, illusions are bound to occur whenever the data are poor or the inferences not appropriate to the logic of a particular situation. Consequently, perceptual knowledge, though based on interpreted experience, is both irredeemably fallible and in principle unrevisable even in the light of conflicting experience. In its *construction* the *finite mind* can play no part, neither through conscious activity nor through blind instinct. This pithy statement happily summarizes

---

1 "[D]es raisonnements instantanés" (ibid.).

2 "[T]ous formés par un acte éternel" (ibid.).

3 "On sentira la main du Tout-Puissant et les profondeur [sic] impénétrables de sa sagesse dans la Providence." (ibid.).

4 "Je dis Dieu et non pas la nature; car ce terme vague de *nature*, si fort en usage, n'est pas plus propre à exprimer distinctement ce qu'on pense que l'*entelechie* d'Aristote." (ibid.).

the double aporia generated by the mechanicist research program as well as the respective solutions suggested by Malebranche.

Notice, however, how in Malebranche's theory of divine intervention the computer analogy is strikingly manifest:

> I believe I should warn again that it is not our soul which forms judgments of distance, magnitude, etc.,... but God does so, in accordance with the laws of the union of mind and body.... However, God makes these judgments in us and for us in exactly the same way as we would form them in ourselves if only we had perfect knowledge of optics and geometry (such as God has); and if we also knew perfectly everything that actually occurs in our eyes and brain; and if our soul could act accordingly in itself, giving itself its sensations. That's why I attribute to the soul the capacity to form these judgments and reasonings and to generate in itself corresponding sensations, which, in point of fact, can only be the effect of an infinite intellect and power. Thus, the very moment we open our eyes, God alone can instruct us instantaneously about the magnitude, the shape, the movement and the colors of the objects surrounding us.[1]

Elsewhere Malebranche goes on to argue (defending his computer analogy rather like a dexterous IBM salesman):

> But since we are not made to occupy ourselves with sensible objects, and to work only for our survival, He saves us that trouble and teaches us instantaneously in a fast and agreeable way a detail as an infinity of truths and miracles.[2]

The information-theoretical problem inherent in the mind-body relationship and rightly exposed by Malebranche he thus resolved by an impossible compromise. For on the one hand perceptions are quite "natural", existing in us, finite minds, simply as "compound sensations" in whose formation we take no part. In so far, then, visual perception (whether compound or not) must be considered as pure *seeing*, as in Descartes. But on the other hand, with respect to God, the Almighty computer, perceptions must be regarded as genuine

---

1 Ibid. Cf. also *Dernier Eclaircissement*, 26; in OM vol. III, G. Rodis-Lewis (ed.), (Paris, 1964), p. 327.

2 "Mais comme nous ne sommes pas faits pour nous occuper des objets sensibles, et pour ne travailler qu'à la conservation de notre vie, il [i.e. God] nous épargne tout ce travail, et nous apprend par une voie abregée et fort agréable en un moment comme infini de véritez [sic] et de merveilles." (*Réponse à M. Regis*, I, 10; OM XVII-1, p. 269).

judgments. For God *interprets* to us our sensory states and accordingly forms the sensations we have. The psycho-physiological gap exposed by mechanicism but which it cannot close, is now bridged by the fiction of a divine computer.

That a primitive computer model is clearly intended also transpires from the fact that God does not act arbitrarily but strictly obeys, like a programmed machine, the fixed laws governing the union of body and mind.[1] This is why distance can only cause an increase in apparent magnitude when registered on the sensory apparatus. Intellectual knowledge, such as an astronomer's knowledge of the real distance of celestial bodies, does not affect the senses and *hence* cannot exert any influence upon our sensations.

Indeed, in sharpening Descartes' theory Malebranche also radicalized the Cartesian dualism of sense and thought. For in Descartes' undeveloped theory of information correlation the fourth means of visual distance discrimination contains a hidden experiential factor. This cue utilized collateral information about the actual dimensions of seen objects, thereby facilitating the visual perception of their real distance. But in Malebranche's theory experience cannot play such a role at all. For knowledge derived from past experience produces no *sensory* data, and thus no input for the God-computer. Consequently, Malebranche does not explain the perception of the unknown differential distance between familiar objects (e.g., a dwarf and a giant three times as large) on the basis of their known differential magnitude. But, conversely, on the basis of their known differential distance (computable from *sensory* data such as accommodation or interposition) he accounts for the perception of their differential magnitude (notwithstanding the equality of their respective retinal images). This explanation makes no use whatsoever of so-called 'empirical' (i.e., derived from experience) or intellectual factors. This is the heart of Malebranche's occasionalism: only changes that are physiologically imprinted upon the senses can occasion God's incredible production of exact psychic responses. These responses are on the whole very dependable for practical life. But from a theoretical point of view they are highly unreliable. Not because the 'computer' might err for all we know, but because its data are physiologically determined and thus valid only relative to the human body. Thus not only is the colorful phenomenological world not a theoretically reliable reflection of the real world which is strictly mechanically defined. But even its apparent geometry often deviates considerably from the real spatial relations of the external world. In both cases objective knowledge of reality as it exists independently

1 Cf. RV, p. 24.

of perception is only possible, according to Malebranche, by means of theoretical science. Experience is an unreliable compass of human understanding.

Thus, despite Malebranche's apparently 'activist' notion of perception as involving interpretation imposed upon arbitrary intelligible signs—an idea which would come to full fruition in Berkeley's theory of perception but which, inchoately, begins its life in Malebranche's pages—the ultimate effect of Malebranche's theory, as in Descartes', is nevertheless to equate perception again with pure sensibility and with passive receptivity of ideas directly triggered by physiological impressions, and to regard perceptual judgment as a mere mode of sensation depending on God's incessant and beneficent interventions. The revolutionary impact of the theory that all perceptual knowledge is based upon *interpreted* experience is thus dulled by the corollary thesis that if there is more to seeing than meets the eye, it is God's Providence alone which, by an eternal act of his benevolence, takes care of these theoretical elements. To us, passive percipients, they are, however, indistinguishable parts of the given-in-experience and thus not modifiable in the light of new experience.

7. Conclusion

Descartes' mechanicist research program comprised a negative heuristic according to which reality was subdivided into exactly two substantial domains, the world of mind and the world of matter. Mind he defined purely intuitionally, matter purely mechanically. Consequently, any scientific account of corporeal states and interactions had to be formulated in purely mechanical and mathematical terms. This even held for physiology and ethology. To appeal to any other ordering principles such as proclaimed by the Aristotelian pseudo-animistic doctrine of souls would be a methodological sin incompatible with the Cartesian program.

Now to bridge the gap between being and thinking, or between body and mind, inevitably called for an information theory of perception. The pattern of neurophysiological stimuli on the walls of the pineal gland is qualitatively entirely unlike the mechanical properties of the external world nor does it bear any qualitative resemblance to the phenomenal and geometrical aspects of the world as perceived. Consequently, that pattern, far from being a qualitatively adequate representation of the real world, can only be regarded as a mere symbolic representation which, though arbitrary, is nevertheless *rich enough from an information-theoretical point of view* to account for the diversity of the perceived qualities of the phenomenal world (i.e., of the world-as-given-in-immediate-perception). This is the *positive heuristic* of the Cartesian program, which, though refined and developed, still clearly inspires the dynamics of Malebranche's enquiries as well. For even though Malebranche required divine agency to deal with the extraordinary complexity of the information-handling processes, yet inasmuch as God only acts on the occasion of physiological impressions one still ought to abide (such is his implicit view) by the Cartesian heuristic of tracing all phenomenal variety to known physiological variety:

> But since He only does it [i.e., process information] as a consequence of the impressions made by those objects upon our body, the reason of the variety of our sensations must be traced to the known variety of those impressions, such as I have tried to do.[1]

However, this positive heuristic in its turn is conceptually constrained by a consequence of the negative heuristic of the mechanicist research program, viz., the *epistemological dualism of sensing and thinking* which parallels the ontological dualism of body and mind. This is why the ingenious information processing so keenly analyzed by Malebranche cannot, according to him, be interpreted as consisting of supra-individual processes. However, if we were to replace the fictitious God-computer by subconscious automatic processes, this would suggest a highly advanced model of the mind comparable to simulation models familiar in contemporary artificial intelligence research. But, needless to say, this theoretical vista is still a far cry from the conceptual world of Malebranche.

Thus, because Malebranche identifies the cognitive with the active and the conscious his program (and that of the Cartesians in general) leaves no room for the development of an autonomous empirical psychology of perception in addition to the mechanical physics and the rational psychology of Descartes. Consequently, Malebranche is very vague concerning such pertinent questions as to what extent the individual has to be aware of the media ("moyens") of distance and magnitude. Again, it is hard to see what role can be played by

---

1 "Mais comme il ne le fait qu'en conséquence des impressions que ces objets font sur notre corps, il faut tirer de la variété connue de ces impressions la raison de la variété de nos sensations, ainsi que j'ai tâché de faire." (RV, p. 24).

individual experiences in the development of mature perception. But the most important consequence of his epistemological dualism is that all the very diverse elements in perception are reduced to the same level as the purely phenomenal qualities such as color or pain. In other words, in spite of the promising theoretical break-through in the realization of complex intellectual contributions being made to ordinary experience, perception in Malebranche is ultimately collapsed back into the realm of pure sensation.

Only in the course of time a gradual sharpening was to take place of the distinction between sensation (or the immediately given in experience) and perception itself. However, it should be clear by now why such a step would not be an easy one to take. For in the first place it would imply the abandonment of the Cartesian dualism of thought and sense. Secondly, it would involve a considerable broadening of the conception of the mind by the gradual, if laborious, recognition of purely cognitive processes being operative at a subconscious level. And finally, as a result of this latter development, it would lead to an extension of the domain of cognitive psychology and to its establishment as an *empirical* science.

In due time the given-in-visual-perception was strictly confined to the retinal impression. Molyneux already initiated this development by observing that the retinal 'image' of distance is a mere point so that distance itself is "invisible".

> For distance of it self, is not to be perceived; for 'tis a line (or a length) presented to our eye with its end toward us, which must therefore be only a *point*, and that is *invisible*.[1]

In Thomas Reid, one century after Malebranche, this led to the more sophisticated conclusion that at any rate there is no *retinal* basis for visual distance discrimination so that, at least, distance is not *immediately* given in vision and thus does not properly belong to the things we 'see'. We will remember that Malebranche, by contrast, also included the decoding results produced by natural judgments as belonging in the last resort to the 'given' in visual perception.

Complementary to this sharpening of the distinction between seeing and judging, or between sensation and perception, more research was being devoted to, and further theoretical development was achieved of, those 'deeper' information-handling processes to be distinguished from the surface processing defining no more than the 'raw data' of vision. As stated above, this development would lead, slowly but surely, to a recognition of subconscious

---

1 Molyneux, *Dioptrics Nova: A Treatise of Dioptrick*, (London, 1692), p. 113.

cognitive processes shaping the phenomenal world. At the same time, however, this would also toll the knell of the Cartesian foundations of the mechanicist program of philosophical psychology. It was not until the mid-19th century that this step could finally be taken. And it was Helmholtz, in particular, who boldly initiated the new research program into the theory of perception, thereby ushering in a new era in the history of psychology.

VII

EPISTEMOLOGICAL ISSUES UNDERLYING THE NINETEENTH CENTURY CONTROVERSIES IN PHYSIOLOGICAL OPTICS. THE HELMHOLTZIAN PROGRAM.

1. The eighteenth century. Rationalist and empiricist developments. Cross-fertilization of originally competing programs.

I have discussed Malebranche's theory in depth as it highlights the fundamental theoretical problem faced by the representationist direction of research into the theory of perception adopted by rationalists and empiricists alike ever since the rise of Cartesian mechanicism and of the corpuscular philosophy in general. We are now in a position to telescope the argument and to deal with relevant contemporary and subsequent 18th century developments in summary fashion.

We have seen that Malebranche, despite his acute interest in analyzing quasi-intellectual processes dealing with subconscious sensory data, yet failed (just as Descartes) to develop a realistic learning or information processing theory. Instead he collapsed these processes and their results into mere sensation, ascribing them to the intervention of God and showing their liability to error, thus underscoring his favorite philosophical thesis that reliable knowledge is to be obtained through thought and intuition alone, while the senses generally deceive.

The empiricists, by contrast, did develop learning theories, but these tended to be purely mechanical (thus Hume's analogy between the principles of gravitation and of association) and they were allowed to act on conscious data alone (thus Berkeley's principle that an idea not itself perceived cannot be the means for the perception of another idea).

Tracing the developments of these respective epistemological (and psychological) versions of the representationist research program into the

18th century we will find that by that time the limitations (some contemporary thinkers even decry them as the "absurd consequences") of the respective programs were most keenly felt in each of their immediate spheres of influence respectively. This gave rise to the peculiar phenomenon of cross-fertilization of originally competing programs. To put it crudely, Condillac (and the French sensationalists in general) inspired by La Mettrie[1] combined the Cartesian doctrine of bio-automatism and the Lockean idea of the soul as a *tabula rasa* on which experience writes all its marks for the elaboration of a mechanical psychology in which not only the more complex experiences but also the most sophisticated cognitive functions and psychological faculties were gradually built up from the most simple experiences under conditions of prolonged environmental exposure. On the other hand, Reid, dissatisfied with what he recognized as the skeptical consequences of the "Ideal System" (if you begin with Descartes' and Locke's theory of ideas you are philosophically destined to end up with Berkeley's and Hume's ridiculous contradictions of common sense), exploited Berkeley's theory of sensory signs for a theory of "natural suggestion" which allowed *sensations* to be like purely immanent (Berkeleyan) ideas but which stipulated that the concomitant *perceptions* which were immediately and naturally suggested by them included objective reference to external objects as well as full-fledged concepts thereof. What was for Hume a product of the imagination was in Reid a product of natural suggestion with immediate evidence and, in view of this, with high initial plausibility. The latter part of the theory obviously harked back to Descartes (the line of influence very probably ran via Fénelon[2]) although the original doctrine had undergone progressive development (e.g., in Reid's theory of justification which allowed concepts and beliefs to be self-evident by our constitution without being infallible for all that).

When in the 17th century it was realized that stimulus and experience could no longer be supposed to be formally identical philosophers (especially empiricists) hoped to re-establish the link between the *ordo essendi* and the *ordo cognoscendi* by assuming instead some kind of *partial* resemblance between phenomenal and stimulus properties. Locke, in particular, made much hay of the notion that the ideas of primary qualities at any rate

---

1 Cf. Julien Offray de La Mettrie, *Histoire naturelle de l'âme* (also known as *Traité de l'âme*), (The Hague, 1745). Also cf. F.A. Lange, *The History of Materialism*, 1st German ed. 1865, tr. by E.C. Thomas, with an introduction by B. Russell (London, 1925), II, 49. For the direction of influence, cf. ibid. pp. 52-3.

2 Cf. L.M. Lacoste, 'La notion d'évidence et le sens commun: Fénelon et Reid', *J. Hist. Phil.*, 15, 3 (July 1977), 293-307.

faithfully reflected corresponding qualities in the external objects. However, since Berkeley called into question the very possibility of a similarity relation between mental and non-mental entities this precious distinction so dear to Locke between primary and secondary qualities was undermined forever.

Closely connected, of course, with this theory of partial resemblance were specific (empiricist) theories of concept formation and of meaning. If such theories originally tended to confirm epistemological realism (e.g., in Locke), in Berkeley (and to a certain extent in Hume) their unexpected idealist potentialities came to the fore. Therefore, whoever wanted to refute Berkeley and Hume was under a clear obligation to suggest a better theory of concept formation than these philosophers entertained. But whoever wanted to develop a new theory of conceptualization not based upon experience-as-a-mirror-of-reality would have great difficulties to avoid ending up either as a dogmatic realist à la Reid or else as a quasi-idealist à la Kant.

2. The Helmholtzian research program into the theory of perception. The true logic of discovery revealed by rational reconstruction of the grand movement of intellectual history rather than by 'faithful' intellectual biographies.

At this point we have reached the epistemological problem facing the 19th century and constituting the (latent) background of various theoretical controversies in physiological optics at the time. And it is to the solution of this problem to which the epistemology as well as the widely ranging scientific research of one particular man addresses itself. That man was Hermann von Helmholtz. Given the foregoing analysis we are now in a position to evaluate the originality of his theoretical contributions. In fact his work amounts to the foundation of an entirely new research program whose epistemological upshot is the vindication of a pragmatic form of hypothetical realism. In contrast to the static and passive theories of perception formulated in the past Helmholtz develops an 'activist' theory according to which even the formal structures of perception are (*pace* Kant) capable of gradual and adaptive development. The theory of mind postulated by this truly information-theoretical account of perception contains elements going far beyond the traditional alternatives of empiricism and rationalism. I will argue that the philosophical literature of early German Romanticism, especially Herder, Schelling, Goethe and Fichte with whose works Helmholtz was demonstrably familiar, provides surprising but highly relevant clues concerning the origin and the latent core of the Helmholtzian research program.

In view of current interpretations of Helmholtz's philosophical position this will no doubt be felt to be a controversial thesis. However, standard interpretations have relied too much on partial analyses of Helmholtz's work as well as on Helmholtz's own understanding of it. Thus they have achieved an air of artificial clarity to the detriment of historical adequacy. Historians have comfortably labeled Helmholtz a bold Kantian[1], a staunch materialist[2], a shame-faced materialist[3], a positivist in disguise[4], or a run-of-the-Mill (so to speak) empiricist[5] according to the particular fragments of Helmholtz's work that they happened to favor. However, if we are to grasp the *total* intellectual profile embodied in Helmholtz's apparently disparate research we must look beyond such limited historical resources as supplied by partial analyses or self-professed philosophical allegiances. Only a rational reconstruction, such as I have attempted to give, of the 'grand movement' of intellectual history regarding the theoretical problems Helmholtz was mainly concerned with may enable us to fathom the intrinsic logic of the problem situation facing him and constituting the principal motive of his work but which he himself, let alone more poorly equipped would-be intellectual biographers, may never have fully realized or uniformly expressed. The extensive introductory chapters of this book, focusing as it does on the historical roots of cognitive science and especially on the significance of Helmholtz's pioneering theory of knowledge and of perception, as well as the historical research these various chapters required were aimed at providing precisely such a historico-structural understanding and at unraveling not the whims of individual scientists but the retroductive logic that implicitly guided their research and the covert rational heuristic that dictated their philosophical positions. If my admittedly ambitious attempt at a rational reconstruction of the theory of perception prior to Helmholtz has served to shed new light on the total intellectual profile underlying Helmholtz's work and to render his

1 Ernst Cassirer, *Das Erkenntnisproblem in der Philosophie und Wissenschaft der neueren Zeit*, vol. IV (Stuttgart, 1957), p. 11.

2 P. Cranefield, 'The Organic Physics of 1847 and the Biophysics of Today', *J. Hist. Med.*, 12, (1957), pp. 407-23.

3 V.I. Lenin, *Materialism and Empirio-Criticism*, A. Fineberg (tr.), (Moscow, 1962), ch. 4, sec. 6.

4 M. Mandelbaum, *History, Man and Reason*, (Baltimore, 1971), ch. 14, sec. 1.

5 Cf. Victor Lenzen, 'Helmholtz's Theory of Knowledge', in *Studies and Essays offered to George Sarton*, M.F. Ashley Montagu (ed.), (New York, 1947), pp. 301-19.

research intelligible *in toto* rather than as a patchwork determined by historical contingencies, then, I trust, this modest merit may mollify the reader's likely exasperation of having been subjected to such lengthy preliminaries.

My analysis so far has shown that the positive heuristic of the representationist research program lacked precisely what the terms of its narrowly mechanistic negative heuristic failed to supply: a dynamic conception of the mind creatively ordering meaningless sensory material at a subliminal level of intelligent interpretive activity. Consequently, the tendency initiated by the positive heuristic to develop a *cognitive* theory of perceptual information processing was again aborted halfway. Thus, content was supposed to be organically imposed, be it "ordained by Nature" (as in Descartes), or by divine intelligence (as in Malebranche), or by principles of natural suggestion operating in virtue of our constitution (as in Reid). Alternatively, perceptual content was regarded as imposed not from within but from without through continued experience resulting in mechanically compounded summations of sensations representing external objects (as in the British empiricists, in Hartley and Priestley, and especially in Condillac). In either case the mind imposes neither order nor interpretation. It is the merely passive recipient of antecedently ordered sensory material.

Both types of account, however, failed to solve the original problem they were designed to solve. That problem was forced upon epistemology ever since the breakdown of naïve Peripatetic realism. It concerned the accessibility of the external world and the objective validity of the concepts used in describing it. The rationalist path, however, ended either in skepticism with regard to the senses (as in Malebranche), or in common-sense dogmatism (as in Reid) justifying the plausibility of perceptual data and concepts by an appeal to the benevolence of the Creator. Alternatively, the empiricist route fared no better. It led either to idealism (as in Berkeley) where the skeptical thrust of Bayle's and Malebranche's arguments was blunted by a retreat into the absolute certainty of purely immanent ideas[1] or else it led to Humean naturalism forfeiting the hope of ever recovering the real world by rational means.

---

1 Baxter concludes ironically that Berkeley's remedy is worse than the disease. Of Berkeley's claim to have refuted skepticism he says:

> This is, I think, as if one should advance, that the best way for a woman to silence those who may attack her reputation, is to turn a common prostitute. He [Berkeley] puts us in a way of denying all things, that we may get rid of the absurdity of those who deny some things. (Baxter, *Enquiry into the Nature of the Human Soul*, vol. ii, p. 284).

To be sure, the first step towards a recognition of the mind's constituent role in perception was taken by Kant. But Kant's conception of the mind's constructive role remained purely formal. In fact, although the mind's operations concerning the data of sense are certainly constructive, they do not constitute genuinely free rational activities. Rather, the mind interprets the data supplied by '*Anschauung*' through the necessary automatic application of a priori categories. Thus the *content* of perception is not in any way affected by the mind's constructive operations. In this sense the old Cartesian dualism of sensing and judging, though considerably modified, is yet retained after all in the Kantian dualism of thought and intuition.

Secondly, the genesis of the formal structures themselves cannot, on Kant's theory, be scrutinized as they have no basis in matters of fact nor in experience but, conversely, condition the latter. Helmholtz, however, in protesting against this Kantian prohibition while embarking upon investigations into the foundations of geometry, of arithmetic and of measurement theory, is animated by his fundamental objection to any theory of knowledge positing ultimate barriers standing between man and reality. Thus he sought to replace Kant's transcendental realism—which in fact represents no more than a constipated form of idealism—by a brand of hypothetical realism. On this Helmholtzian view the mind acts *dynamically*, building up a conjectural world gradually refined by small leaps of analogy[1] and constantly checking its interpretations of the data of sense through the ongoing interplay of volition, sensation and reflection.

3. The relevance of German Romanticism to the Helmholtzian program.

This novel conception of the creative mind, I maintain, was prepared in early German Romanticism and derived not from Kant but from Fichte. Furthermore, wherever in the course of the 19th century the Kantian heritage

---

1 Hermann von Helmholtz, *Handbuch der Physiologischen Optik*, 3 vols. (Heidelberg, 1856/1860/1867), 875 pp.; 2nd ed., in 8 parts (Hamburg, 1885-1894) 640 pp.; 3d ed., 3 vols. (Hamburg, 1909-1911). Eng. tr. by James P.C. Southall, *Helmholtz's Treatise on Physiological Optics*, 3 vols., from the 3rd German ed., published by the Optical Society of America (Rochester: New York, 1924-1925; Dover reprint, 3 vols. in 2: New York, 1962), 482, 479 and 746 pp., vol. III, p. 4. The Southall edition will be referred to below as PO, all other editions as PO followed by the year of publication.

appears to loom large it is important to bear in mind that the epistemological mainstream tended to be Kant as understood and disseminated by the Hegelians who dominated university posts in philosophy.[1] It would be perverse, of course, to suggest that Helmholtz, and with him the naturalistic anti-Hegelian movement that flourished after the collapse of German Idealism and of which he was a chief exponent, was in fact directly influenced by Hegel's or Schelling's ideas. In scientific circles the idealist philosophers were mainly remembered as the masterminds of an apriori *Naturphilosophie* which the later scientific community could not help but condemn as the conceited products of an overweening philosophical ambition.[2] But it would be equally naïve simply to take Helmholtz's word for what he is doing and to regard him as a pure representative of the Kantian creed.

To be sure, Helmholtz did lead the 'back-to-Kant' movement in the mid-19th century.[3] However, at a time that among fellow scientists philosophy in general, because of its easy identification with the erratic fancies of the idealist armchair philosophers, was in utter disrepute[4] this proclaimed philosophical

---

1 Cf. Eden and Cedar Paul (trs.), *Treitschke's History of Germany in the Nineteenth Century*, 7 vols., (London, 1919), IV, 568-9; Rudolf Virchow's 1893 address on 'Die Gründung der Berliner Universität und der Uebergang aus dem philosophischen in das naturwissenschaftliche Zeitalter', in *Idee und Wirklichkeit einer Universität. Documente zur Geschichte der Friedrich-Wilhelms-Universität zu Berlin*, Wilhelm Weischedel (ed.), 2 vols., (Berlin, 1960), II, p. 420; Charles A. Culotta, 'German Biophysics, Objective Knowledge, and Romanticism', in *Historical Studies in the Physical Sciences*, Russell McCormmach (ed.), 4, (1974), pp. 3-38; Yehuda Elkana, 'The Problem of Knowledge in Historical Perspective', in *Proceedings of the 2nd International Humanistic Symposium*, (Athens, 1974), p. 236.

2 Hegel's dissertation, for example, ridiculed the search for new planets after Ceres had just been discovered but before the news of this event had spread to Germany. Duke Ernest of Gotha sent the dissertation to the astronomer Zach with the superscription "Monumentum Insaniae saeculi decimi noni"; cf. R. Wolf, *Geschichte der Astronomie*, (München, 1877), pp. 684 ff.

3 Cf. Theobald Ziegler, *Die geistigen und sozialen Strömungen des neunzehnten Jahrhunderts*, (Berlin, 1911), p. 317; Ernst Cassirer (1957), p. 11; Thomas E. Willey, *Back to Kant*, (Detroit, 1978), p. 26.

4 In 1847 the current was so much against philosophy in general that the 25-year old Helmholtz was persuaded by his friend DuBoys-Reymond to drop his philosophical introduction to his celebrated 'Ueber die Erhaltung der Kraft' as it might jeopardize the publication not only of the introduction itself but of his scientific paper as well. In fact, it *was* rejected by the narrow-minded anti-speculative empiricist Poggendorff who thought the subject matter not sufficiently experimental to justify publication in the

allegiance is hardly surprising. Almost any philosophy claiming to offer more than pure critical epistemology would have fallen on deaf ears. Secondly, Kant's epistemology spelled out in astonishing detail the formal conditions under which according to him human perception and cognition must occur. Inasmuch as his transcendental account could be re-interpreted as an elaborate theory of the necessary participation of the human mind and its psycho-physiological structures in all epistemic processes and their effective contribution to all our knowledge of reality Kant's epistemology was bound to be congenial and stimulating to the 19th century sense-physiologists who shared a similar theoretical concern. Not surprisingly Rokitansky called Kant "der Mann der Physiologen".[1] But, of course, merely to share some guiding ideas that only serve to define in very general terms a common *problem* is not enough to establish an alleged continuity of (epistemological) research programs. In order that this be the case one must also share a heuristic which serves to anticipate a common *solution*. I will argue that this latter condition is not fulfilled in the philosophical relation between Helmholtz and Kant.

Quite a few Helmholtz interpreters have similarly detected heterodox elements in Helmholtz's self-professed Kantianism. Neo-Kantians especially were not a little disturbed by what they regarded as an intolerable tension between Helmholtz's philosophical claims and his theoretical heresies.[2] Again, apparently following Helmholtz's lead they now declared him to be an empiricist. But what is an empiricist? Someone who believes that knowledge must have a basis in experience? But Kant believed that too, and so did the positivists who would reject Helmholtz's realism. Even Descartes would not deny this nor, as I have explained above, would his denial be inconsistent with his theory of method, as is widely believed. Perhaps, then, empiricism is the belief that experience is the only basis of (empirical) knowledge? But this even Locke did not believe[3] nor did the naïve Peripatetic

---

Poggendorff's Annalen [cf. Leo Koenigsberger, *Hermann von Helmholtz*, F.A. Welby (tr.), (Oxford, 1906), p. 38].

1 Cf. A. Riehl, *Führende Denker und Forscher*, (Leipzig, 1924), p. 225.

2 Cf. J. Schwertschlager, *Kant und Helmholtz erkenntnistheoretisch verglichen*, (Freiburg-im-Breisgau,1883); Alois Riehl, 'Helmholtz in seinem Verhältnis zu Kant', *Kant-Studien*, 9 (1904), pp. 261-85; Albrecht Krause, *Kant und Helmholtz: Ueber den Ursprung und die Bedeutung der Raumanschauung und der Geometrischen Axiome*, (Lahr, 1878); J.P. Land, 'Kant's Space and Modern Mathematics', *Mind* 2 (1877), pp. 38-46.

3 Cf. Thomas E. Webb, *The Intellectualism of Locke*, (Dublin, 1857); Maurice Mandelbaum, 'Locke's Realism', in id., *Philosophy, Science, and Sense Perception*, (Baltimore, 1964), pp. 1-60.

empiricists whose well-known dictum that nothing is in the mind that was not first in the senses Locke adopted as the motto of the empiricist tradition.

Clearly, such tidy but tedious dichotomies as reason *vs.* experience or Kant *vs.* empiricism are generally very hard to sustain in historical detail. But they altogether fail to do justice to the rich complexity of 19th century thought. I believe that what is innovative in Helmholtz's empiricism and heretic in his Kantianism is most readily understood if we take more seriously the idea—itself a sound insight of 19th century thought—that history in general, and especially the history of ideas, is a continuous development through oppositions and antagonisms. What Helmholtz most ardently sought, i.e., to vindicate realism within a representationist context, required a radically new conception of the mind, the material for which was supplied to him by the romantic reactions to the Enlightenment in Germany which included a reorientation toward non-Lockean theories of knowledge based upon organismic conceptions of mental processes and a quest for a learning theory and psychology that would stress the *creative* impulse of the individual mind rather than its passive capacity to receive images or information already processed. Indeed, such sentiments were not confined to German artists and intellectuals. Thus, Wordsworth, the leader of the romantic movement in England, spoke of the mind as a lamp, not as a mirror.[1] Now Helmholtz's mind was partly nurtured in the romantic era. Fichte was the philosopher his father most favored and respected. Helmholtz would later quote him frequently and almost without exception approvingly. Fichte's son was a close friend of the Helmholtz family. Finally, Goethe's creative genius in science, in art and philosophy sufficiently impressed Helmholtz to devote two important public addresses entirely to Goethe's memory[2] in which he noted intimate epistemological similarities between the poet's views and his own.[3]

---

1 Cf. M. Abrams, *The Mirror and the Lamp, Romantic Theory and the Critical Tradition*, (New York, 1958). For similar views, cf. Paul Kluckhorn, *Die Idee des Menschen in der Goethezeit*, (Stuttgart, 1946), esp. pp. 33-8 on Fichte, Schlegel, Schelling, and Carus.

2 *'Ueber Goethe's naturwissenschaftlichen Arbeiten'*, (1853), in Hermann von Helmholtz, *Vorträge und Reden*, 2 vols. (5th ed., Braunschweig, 1903), vol. I, pp. 23-47; and 'Goethe's Vorahnungen kommender naturwissenschaftlicher Ideen', (1892), in ibid. II, pp. 335-61. This collection of essays and addresses will be referred to below as VR.

3 Cf. VR II 358-9.

4. Helmholtz's theory of subliminal cognitive activity.

Throughout his life Helmholtz has been fascinated by the creative resources of man's perceptual powers. He was irresistibly drawn towards a study of the more primitive aspects of cognitive life. He thus pursued lines of investigation which *in opposition to* Kant had been urged first by Herder and later by the romantic and idealist philosophers of the early 19th century.[1] However, rather than romanticizing or mystifying the divine inspiration of the poet or the aperçus of the artist-scientist as involving inexplicable divinations (as in Jacobi) or mere leaps of faith into the unknown without any intelligible connection with the known and the acquired (as in Fichte[2]), Helmholtz was convinced that intellectual processes not unlike our conscious thoughts underlie our boldest discoveries and our most creative insights. The popular philosophical contrast between thought and intuition inherited from the Cartesian dualism between thinking and sensing he expressly denied.[3] Kant's theory of intuition he deemed woefully inadequate.[4] He draws an analogy between artistic genius and ordinary perception. In fact, artistic imagination and perceptual intuition are sprouts from the same seed, the one more matured and more refined than the other but both essentially alike and nurtured by the same soil: experience.

> The very same considerations which I have here made to bear first on the example of sensory perceptions, are also entirely applicable to artistic perceptions. To be sure, these arise effortlessly, with sudden clarity, without the owner knowing whence they came. But it absolutely doesn't follow that they shouldn't contain results derived from experience, incorporating the accumulated memories of their lawlike character.[5]

---

1 Thus, according to Hegel, intelligence is not primarily receptive (*aufnehmend*) but essentially active (*thätig*) and it is inseparably connected with the will. In fact, all so-called faculties are internally related, but Hegel especially insists on the interpenetration of thought and will. This, of course, directly contradicts the Cartesian theory which opposed the merely contemplative intellect to the active and often intruding will. Cf. Hegel's *Encyclopaedia of the Philosophical Sciences*, in *Sämtliche Werke*, Hermann Glockner (ed.), 20 vols. (Stuttgart, 1928) VI, 260-72, (sec. 368-87).

2 Cf. Johann Gottlieb Fichte, *Popular Works*, (London: Trübner, 1889), I, 411.

3 VR II 341.

4 VR II 338.

5 "Dieselbe Ueberlegungen die ich hier zunächst an dem Beispiel der sinnlichen Anschauungen angestellt habe, lassen sich vollständig auch auf künstlerischen Anschau-

But 'truth' in art and in perception is attainable neither by the sheer mechanical impact of accumulated experiences upon the mind as the British associationists and the French sensationalists once thought, nor on the other hand by the undirected play of free association and of unleashed imagination as the poets of the romantic area fancied. Thus Helmholtz continues the above quotation as follows:

> This opens up a positive source of artistic imagination which is entirely suitable to explain the rigorous consistency of great works of art, in contrast to the free play of the fantasy once so celebrated by the poets of the Romantic Tradition.[1]

Thus Helmholtz provides the framework for a theory of perception which is in close agreement with the main tenets of the more promising developments of modern research into the area of pattern recognition. The perceptual problems posed to the visual system are characterized by a solution which is the more narrowly definable the richer our stock of relevant experiences and the more intelligent their organization.[2] What is required for their solution is creative rather than merely novel ideas. Creative ideas are novel ideas that are highly adaptive. They are produced more effectively by directed than by undirected thought.[3]

---

ungen übertragen. Daraus, dass sie mühelos kommen, plötzlich aufblicken, dass der Besitzer nicht weiss, woher sie ihm gekommen sind, folgt durchaus nicht, dass sie keine Ergebnisse erhalten sollten, die aus der Erfahrung entnommen sind, und gesammelte Erinnerungen an deren Gesetzmässigkeit umfassen." (VR II 344).

1 "Hierdurch werden wir auf eine positive Quelle der künstlerischen Einbildungskraft hingewiesen, welche auch vollständig geeignet ist, die strenge Folgerichtigkeit der grossen Kunstwerke zu rechtfertigen, im Gegensatz zu dem einst von den Dichtern der Romantischen Schule so gefeierten freien Spiele der Phantasie." (ibid.).

2 Cf. R.L. Gregory's computer programs designed to give 'scene analysis' (recognizing objects from pictures by computer) by means of conditional probabilities giving rise to interactions which may generate visual effects; in R.L. Gregory, *Concepts and Mechanisms of Perception*, (New York, 1974), p. XXX. Also cf. id., 'A Speculative Account of Brain Function in terms of Probability and Induction', ibid., pp. 521-36; as well as his stimulating paper entitled 'How So little Information Controls So Much Behaviour', ibid., pp. 589-601. Also cf. I. Rock, 'In Defense of Unconscious Inference', in W. Epstein (ed.), *Stability and Constancy in Visual Perception: Mechanisms and Processes*, (New York, 1977); and id., *The Logic of Perception*, (Cambridge, 1983). More recently, connectionist models of (supervised and unsupervised) learning networks provide promising theoretical developments of a different kind, yet consistent with fundamental Helmholtzian insights; cf. Rumelhart, D.E. & McClelland, J.L. (eds.), *Parallel Distributed Processing, Explorations in the Microstructure of Cognition*, 2 vols., (MIT Press, 1986).

3 Cf. Barry F. Anderson, *Cognitive Psychology*, (New York, 1975), p. 147.

Likewise, the more primitive subconscious operations of cognitive life at the level of both artistic imagination and the formation of ordinary perceptual hypotheses may, according to Helmholtz, be regarded as continuous, in a sense yet to be defined, with the operations of conscious intellectual life. Both activities involve interpretive conjectures guided by a heuristic which is informed and inspired by experience. Clearly, this idea involves a theory of mind and a theory of perception which on the one hand, notwithstanding Helmholtz's frequent appeal to associative processes and in spite of his self-proclaimed empiricist bent, goes far beyond the mechanical learning theories of British empiricism and French sensationalism. On the other hand, notwithstanding its essential reference to computation and to (unconscious) inference, it also goes far beyond the bio-automatism and the nativism inherent in the traditional rationalist research program. This new orientation which in fact defines a new research program into the (naturalistic) theory of (perceptual) knowledge while at the same time opening up an entirely new field of study for cognitive, experimental and developmental psychology, is most simply expressed in Helmholtz's pithy statement:

> The fundamental thesis of the empirical theory is: *The sensations of the senses are tokens for our consciousness, it being left to our intelligence to learn how to comprehend their meaning.*[1]

This apparently simple notion which constitutes the foundation of Helmholtz's physiological inquiries is coupled with the determination to vindicate some form of epistemological realism. Thus Helmholtz chooses to oppose any unrevisability theory variously embodied in theories of common sense, of a priori categories, or of primitive perceptions. This defines the scope and the underlying theme uniting Helmholtz's various topics of research ranging from physiological optics and acoustics to the foundations of mathematics and of measurement theory. This is especially clear from the opening statements to his *Numbering and Measuring from an Epistemological Viewpoint* (1887), where he writes:

> Although numbering and measuring are the foundations of the most fruitful, sure and exact scientific methods known to us at all, relatively little work has been done on their epistemological foundations. On the philosophical side strict disciples of Kant, who adhere to his system exactly as it developed historically amidst the views and knowledge of his time, had of course to regard the axioms of arithmetic as propositions

1 PO III 533.

> given *a priori*, which narrow down the specification of the transcendental intuition of time. ***Through this conception, the issue of a further foundation and derivation for these propositions was terminated in both cases.***
>
> In earlier writings I have endeavored to show that the axioms of geometry are not propositions given *a priori*, but that they are rather to be confirmed and refuted through experience...
>
> It is then clear that if the empiricist theory—which I besides others advocate—regards the axioms of geometry no longer as propositions unprovable and without need of proof, it must also justify itself regarding the origin of the axioms of arithmetic, which are correspondingly related to the form of intuition of time.[1]

5. Helmholtz's research program contrasted with competing epistemological programs.

Now my historico-structural reconstruction is corroborated by the fact that, having thus identified Helmholtz's fundamental epistemological motivation, we will readily recognize that various contemporary philosophical systems radically different from Helmholtz's theoretical position in appearance as well as in substance, yet started from a strikingly similar perception of the epistemological problem at hand. For the attempt to gain access to cognitive sources other than the ones circumscribed by Kant's apriorism was a theme common to most early 19th century thought. It took vastly different forms, however, with varying degrees of success. For convenience sake, we might distinguish the romantic or 'irrational' movement (Jacobi, Fichte), the dialectical or 'rational' movement (Hegel, Schelling), the creative or

1 Hermann von Helmholtz, 'Zählen und Messen erkenntnistheoretisch betrachtet', repr. in id., *Wissenschaftliche Abhandlungen*, 3 vols. (Leipzig, 1882, 1883, 1895), III, pp. 356-91; these works will be referred to below as WA. Also in *Schriften zur Erkenntnistheorie*, ed. with extensive notes by P. Herz and M. Schlick, (Berlin, 1921); tr. by M.F. Lowe as: *Epistemological Writings*, Boston Studies in the Philosophy of Science vol. XXXVII, R.S. Cohen and Y. Elkana (eds.), (Dordrecht and Boston, 1977), p. 72 (emphasis added). The German original of this book will be referred to below as SE, the English translation as EW.

'anarchical' movement (Herder, and the Young Hegelians he influenced), and the naturalistic or 'empirical' movement (Fries, Helmholtz, Zeller, Spencer, Bain).

Herder, in opposition to Kant, had insisted that man *creates* truth. History he therefore regarded as a creative process, a multifarious effervescence of human creativity which is utterly contingent and provides no ultimate solutions. Moreover, just as cultures are incommensurable, so are intellectual visions. Consequently, he had emphasized process, pluralism and freedom, notions which would later inspire the Young Hegelians.

By contrast Hegel, though impressed with Kant, had taken issue with Kant's (as well as Jacobi's and Fichte's) dogmatic "metaphysics of objectivity".[1] Like Jacobi and Fichte, Hegel sought to overcome the limitations Kant had imposed upon the Understanding by claiming a higher authority for Reason. But unlike them, he founded Reason neither on feeling nor on faith. In fact, it was not the cognitive faculty of any individual. Rather, the dialectic of Reason operated on the supra-conscious plane of world history, guiding with rational necessity the development of Mind through its various global stages. Thus, in sharp contrast with Herder, truth was not attainable by any one individual nor expressible by any one culture or by any particular intellectual articulation. On the contrary, history's only subject was Reason, not man, and the rational and the actual simply coincided.

Now we might say, that Helmholtz similarly extends the scope of reason in order to overcome the fixed boundaries set by the Kantian categories and defining the logical limits of human knowledge. However, instead of seeking such an extension on a supra-conscious plane (as Hegel had done), Helmholtz endeavors to extend the operations of reason into the realm of individual subconsciousness. Moreover, the will which in the previous epistemological tradition of pure mental receptivity was banned from the cognitive process Helmholtz now regards as essential to our *individual* learning processes and to our eventual knowledge of the world.[2] Thus the apparently fixed forms exhibited in mature perception and thought are produced by rational development and must be traced to intelligent ontogenetic adaptation as their ultimate ground. Helmholtz might even have agreed with Reid that certain perceptual beliefs are evident without any *conscious* argument or

---

1 G.W.F. Hegel, 'Glauben und Wissen, oder die Reflexionsphilosophie der Subjektivität, in der Vollständigkeit ihrer Formen, als Kantische, Jacobische und Fichtesche Philosophie', in *Sämtliche Werke*, I, 431.

2 VR II 359.

reasoning being offered in their defense. Against Reid, however, he would maintain that this is so because such beliefs are the reliable results of cognitive subconscious operations, of creative mental activity, of *tested* conjectures. Similarly, he could agree with Kant that the forms of perception are a priori (in a sense yet to be defined), and nonetheless insist upon the need to inquire into their epistemological foundations and their possible basis in matters of fact and experience.

This fundamental naturalistic perspective unites practically all of Helmholtz's research while radically separating it from various competing epistemological programs of his time. It is emphatically expressed in the repeated reference to 'facts' in the titles of such prominent papers as his *On the Facts Underlying Geometry* [1] and his *The Facts in Perception.*[2] Finding the universal facts of experience which would also be analytically adequate to serve as a possible foundation of what Kant had regarded as the irreducible forms of perception would (to paraphrase Priestley's criticism of Reid) relieve dame Nature of the *unnecessary* load which Dr. Kant had laid upon her.[3] For it would render such forms not incomprehensible as both the condition and the mechanical product of compounded sensations (as suggested by the traditional empiricist direction of research), nor inexplicable as the irreducible a priori structures constitutive of experience (as in Kant). Rather, in virtue of Helmholtz's novel conception of mental activity, such forms would be rendered intelligible as the products of subconscious inferences and as constructions of space and time which, though conjectured, yet have a solid basis in universally experienced matters of fact.

Helmholtz's central epistemological problem was how to reconcile a truly information-theoretical account of perception with a theory of objective (perceptual) knowledge. What he needed, and formulated, was a master theory. To this end he developed his unusual theory of unconscious inferences. It shows that his 'empiricism' is far from traditional. For his theory holds that the percipient is not stripped of all activity while merely inertly interacting with a surrounding medium of mechanical energies. Rather, the percipient is regarded as an active participant in the perceptual process and as gradually 'tuning in' to the unknown structures of the surrounding world. Thus he can do more than merely absorb whatever comes his way through experience as well as being conditioned by it. He can also

---

1 'Über die Tatsachen, die der Geometrie zu Grunde liegen' (1868).

2 'Die Tatsachen in der Wahrnehmung' (1878).

3 Joseph Priestley, *An Examination of Dr. Reid's Inquiry*, pp. 18 f.

break through the walls of his "intraphenomenal prison"[1] and genuinely *learn* from experience by means of creative activity and intelligent experimentation, thus gradually improving his acquired knowledge by novel experience and tested conjectures. In a sense, then, we might say that Helmholtz's epistemological problem on the level of perception is essentially comparable to Whewell's problem on the level of conscious theory formation.[2]

In the following we will trace the development of Helmholtz's naturalistic perspective and discuss various aspects of his complex relationship to Kant. We will emphasize the interwovenness of Helmholtz's physiological, philosophical, and mathematical investigations. Finally, we will discuss his notorious theory of unconscious inferences and bring out its relevance to modern psychology and philosophy of perception. This theory, perhaps more than everything else in his scientific work, may serve to show that far from being a victim of Victorian empiricism, the boldness of Helmholtz's theorizing rather testifies to the speculative tradition of 19th century German education and to the comprehensive philosophical spirit that characterized its scientific institutions.

---

1 Paul Dubois-Reymond declared these walls "impenetrable" in his 'Über die Grundlagen der Erkenntnis in den exakten Wissenschaften', in which physics is doomed to failure from the beginning in its attempt to grasp reality; cf. S.W. Itzkoff, *Ernst Cassirer: Scientific Knowledge and the Concept of Man*, (Notre Dame, 1975), pp. 45-50. This is in line, of course, with positivist tendencies variously expressed towards the end of the 19th century by such men as Paul's brother Emil DuBois-Reymond, Mach and Spencer. However, to place Helmholtz's philosophical position squarely within the "positivist strand" of 19th century thought, as Mandelbaum does, is to ignore the novelty and the central tenets of his ideas (cf. Maurice Mandelbaum's majestic *History, Man and Reason, A Study in Nineteenth-Century Thought*, (Baltimore, 1971), ch. 14.

2 Cf. Robert E. Butts, 'Whewell's Logic of Induction', *Foundations of Scientific Method: The Nineteenth Century*, R.N. Giere & R.S. Westfall (eds.), (Indiana, 1973), pp. 53-85.

VIII

THE INTERPLAY BETWEEN PHILOSOPHY AND PHYSIOLOGY IN HELMHOLTZ'S VIEW.

1. Helmholtz's Conception of Philosophy in Historical Perspective.

Few nineteenth century scholars have been more avowedly Kantian in rhetoric, yet less Kantian in actuality than Hermann von Helmholtz. The son of a Prussian philologist and Gymnasium teacher at Potsdam, Helmholtz grew up in an intellectually stimulating milieu in which philosophical issues were the objects of keen interest. Yet, during the days of his intellectual maturation, his special appreciation of Kant stemmed from his very opposition to the philosophy of his time.

Born in 1821, Helmholtz's early development took shape in an era that saw the rapid decline of German Idealism.[1] Especially among scientists a growing opposition arose against speculative philosophy. The mechanical conception of the universe had finally found wide-spread acceptance also in Germany. Materialist ideas were in the ascendant.[2] Helmholtz, not particularly

---

1 On the breakdown of German Idealism cf. John Theodore Merz, *A History of European Thought in the Nineteenth Century*, 4 vols. (1904-1912), (reprint: New York, Dover, 1965), III, pp. 177-81; Heinrich Treitschke, *History of Germany in the Nineteenth Century*, E. and C. Paul (trs.), 7 vols. (London, 1919), V, p. 587; Ludwig Büchner, *Sechs Vorlesungen über die Darwin'sche Theorie,...* (Leipzig, 1872), (1st ed., 1868), pp. 29-31.

2 Karl Fortlage, who was far from sympathetic to the materialist movement, ominously summarized the onslaught of materialism in mid-19th century Germany as follows:

> "A new world view is settling itself in the minds of men. It goes about like a virus. Every young mind of the generation now living is affected by it.... Only this much is certain: the old has become obsolete and the new presses powerfully forward, as if mankind were pregnant with a new humanity." ['Materialismus und Spiritualismus', in *Blätter für literarische Unterhaltung*, 30 (1856), p. 541].

bellicose by nature, had nothing but contempt for what he called the *Ikarusflug der Spekulation*. He waged a life-long war against metaphysics and against those philosophers who refused to study facts and yet deemed themselves fit to pronounce on them. He writes to the mathematician Lipschitz in 1881:

> In my mind I scoff like Schopenhauer at the professional philosophers; but I don't want to put it to paper. Everybody only reads his own work and is incapable to assimilate the thoughts of others.... No doubt, in the final analysis false rationalism and theorizing speculation constitute the greatest flaw of our German education in all its forms.[1]

Helmholtz's philosophizing marks an interesting transitional stage in the history of ideas connecting Kant with the modern era. As we shall see, however, those of his views that are at substantial variance with both of these traditions are more worthy of our attention. At any rate, his attack on metaphysics and the strict conception of philosophy he advocated hark back to the spirit of Königsberg while at the same time unmistakably adumbrating some of the main tenets of the Vienna Circle. Yet, in his opposition to 'metaphysics' he was following the wisdom of his day rather than his own deeper philosophical inclinations. The decline of German Idealism had fostered a radical reaction among the German scientific community which tended to replace the older *Naturphilosophie* with no philosophy at all.

Again, Büchner, in 1872, noted the apparent historical significance of his *Kraft und Stoff*, which had "undergone twelve big German editions in the short span of seventeen years. Which further has been issued in non-German countries and languages about fifteen to sixteen times in the same period, and whose appearance (although its author was entirely unknown up to then) has called forth an almost unprecedented storm in the press..." [Ludwig Büchner, *Aus Natur und Wissenschaft*, (Leipzig, 1874), I, p. 3]. Also cf. Frederick Gregory, *Scientific Materialism in Nineteenth Century Germany*, (Dordrecht, 1977), *passim*; Friedrich Albert Lange, *The History of Materialism*, E.C. Thomas (tr.), (London, 1925), Vol. II; Heinrich Treitschke (1919); Owsei Temkin, 'Materialism in French and German Physiology of the Early Nineteenth Century', *Bull. Hist. Medicine*, 20 (1946), pp. 322-7.

1 "In meinen Gedanken schimpfe ich wie Schopenhauer auf die Philosophen von Fach; aber ich will es nicht zu Papier bringen. Jeder liest nur sich selbst und ist unfähig, sich in die Gedanken anderer hineinzudenken.... Schliesslich ist der falsche Rationalismus und die theoretisierende Spekulation doch der schwerste Mangel unserer deutschen Bildung nach allen Richtungen hin." [Quoted in Leo Koenigsberger, *Hermann von Helmholtz*, (Braunschweig, 1902-3), 3 vols., II, pp. 163-4; also in Friedrich Conrat, *Hermann von Helmholtz's Psychologische Anschauungen*, (Halle a.d. S., 1904), pp. 264-5].

Younger experimentalists like Matthias Schleiden, who were extremely critical of the *Naturphilosophie* espoused by Hegel, Schelling and Oken[1] urged that philosophy be abandoned and its problems be tackled by experimental science which had recently proved so highly successful in solving similar 'eternal' riddles of philosophy.[2] Helmholtz, by contrast, was not opposed to philosophy. On the contrary, his was a deep concern for philosophy, but a grudge against it. According to Helmholtz the idealists had corrupted philosophy by trespassing the boundaries of legitimate philosophical inquiry. They espoused wild metaphysical speculations with dogmatic fervor, presented them as more than mere hypotheses, and declared them exempt from scientific scrutiny. The task of the various sciences was usurped by "die Philosophen von Fach" who hardly deigned to even draw inspiration from empirical investigations. All this (in his view) understandably alienated men of science. Now that intellectuals all over Germany were awakening from the rhapsodic trance of the past generation only to find Hegel's and Schelling's shamanistic chants full of empty verbiage and devoid of any noticeable enlightenment, there was a real danger, however, that they might turn their backs not just on German Idealism, but on philosophy itself.

Helmholtz's fear of such a wholesale rejection of philosophy was no less than his professed distaste for the field of metaphysics.

> In thus confining the name of Metaphysics to that so-called science which strives by pure thought to formulate conclusions as to the ultimate principles of the coherency of the Universe, I must protest against my objections to metaphysics being transferred to philosophy in general. In my opinion nothing has been so pernicious to philosophy as its repeated confusion with metaphysics. The latter has played much the same part in relation to the former as that which astrology has borne to astronomy.[3]

---

1 Cf. Matthias Schleiden, *Über Schellings und Hegels Verhältnis zur Naturwissenschaft*, (1844).

2 Cf. Matthias Schleiden, *Über den Materialismus der neueren deutschen Naturwissenschaft, sein Wesen und seine Geschichte*, (Leipzig, 1863), pp. 45-6; Jacob Moleschott, *Ursache und Wirkung in der Lehre vom Leben*, (Giessen, 1867), pp. 3-4; Heinrich Böhmer, *Geschichte der Entwicklung der naturwissenschaftlichen Weltanschauung in Deutschland*, (Gotha, 1872), p. 143.

3 Quoted in Königsberger (1906), p. 427.

The proper task of philosophy, according to him, was "the investigation of epistemological processes and of scientific methods".[1] But here, indeed, it had to play a role of vital importance,

> a task, which will always remain the responsibility of philosophy, and from which no era can withdraw with impunity.[2]

In both these concerns—the desire to rescue epistemological philosophy from the shambles of idealism and to protect it from unwarranted metaphysical intrusions—Helmholtz thought he was advocating no more than a squarely Kantian position. Throughout his life he regarded Kant as the bold anti-metaphysician who had delineated once and for all the proper boundaries of empirical and philosophical inquiry. When he later realized that his own investigations had led him to results which were incompatible with certain Kantian doctrines, he interpreted these deviations as mere extensions and radicalizations of Kant's basic insights.

> In my opinion Kant's great achievement can only be maintained if we drop his error concerning the purely transcendental meaning of the geometrical and mechanical axioms. However, this would at the same time make it impossible to use his system as a foundation of metaphysics. And I suspect that this is the real reason why all those among his supporters who entertain metaphysical hopes and inclinations, attempt to cling to these contested points.[3]

Helmholtz's view of the aims and methods of an epistemologically conceived philosophy, however, differs in at least one important respect both from Kant's critical philosophy, and, for that matter, from the formal philosophy of the ensuing analytic tradition in that he demands explicit recourse to empirical inquiry. The construction of a sound theory of knowledge is a shared concern and a common task of philosophy *and* the natural sciences. Notwithstanding

---

1 In ibid., p. 139.

2 "ein Geschäft, welches immer der Philosophie verbleiben wird, und dem sich kein Zeitalter ungestraft wird entziehen können." ['Über das Sehen des Menschen' (1855), in *Philosophische Vorträge und Aufsätze*, H. Hörz and S. Wollgast (eds.), (Berlin, 1971), p. 47; also VR I 88].

3 "Meines Erachtens kann man, was Kant Grosses geleistet hat, nur halten, wenn man seinen Irrtum über die rein transcendentale Bedeutung der geometrischen und mechanischen Axiome fallen lässt. Damit fällt aber auch jede Möglichkeit, sein System zu einer Grundlage der Metaphysik zu machen und dies scheint mir die innere Grund zu sein weshalb sich unter seinen Anhängern alle die metaphysische Neigungen und Hoffnungen haben, an diese bestrittenen Punkte anzuklammern suchen." [Quoted in Koenigsberger (1902-3), II, p. 142; also in Conrat (1904), pp. 266-7].

Schlick's distressed attempt to offer a more favorable interpretation of the following passage (thus 'exonerating' Helmholtz from the intellectual blemish, otherwise incurred, of having violated the philosophical statutes of Vienna) Helmholtz's conception of epistemology tends emphatically towards a psycho-physiology of perceptual and cognitive processes.

> ...the fundamental problem placed at the beginning of all science was the problem of epistemology: "What is true in our intuition and thought? In what sense do our representations correspond to actuality?" Philosophy and natural science encounter this problem from two opposite sides, it is a task common to both.
>
> The former, which considers the mental side, seeks to separate out from our knowledge and representation what originates from the influences of the corporeal world, in order to set forth unalloyed what belongs to the mind's own workings. Natural science, on the contrary, seeks to separate off that which is definition, symbolism, representational form or hypothesis, in order to have left over unalloyed what belongs to the world of actuality whose laws it seeks. Both seek to execute the same separation, although each is interested in a different part of what is separated. In the theory of sense perceptions, and in investigations into the fundamental principles of geometry, mechanics and physics, even the inquirer into nature cannot evade these questions.[1]

Thus, strictly a priori analyses—whether transcendental analyses of experience and of thought in Kant's manner, or formal analyses of empirical concepts in the fashion of the twentieth century phenomenalists—simply do not suffice to reconstruct the integrated chain of perceptual and cognitive processes and their complicated interactions. And by neglecting the descriptive task the epistemologist leaves his normative program dangling in mid-air as well. For if we are to suppose, as Reichenbach suggested, that the formal epistemologist's theoretical reconstruction is not arbitrary but bound to actual thinking by the "postulate of correspondence"[2], then it becomes hard to see how we can affirm any such correspondence unless we have a relatively clear notion of what exactly are supposed to be reconstructing.

---

1 'The Facts In Perception', (1878), EW 117-8. For Schlick's mopping-up exercise cf. ibid., pp. 163-4, n. 4.

2 Hans Reichenbach, *Experience and Prediction*, (Chicago, 1938), pp. 5-6.

In fact, the problem is more pernicious than this general argument alone suggests. For it is a basic assumption of the formal epistemologist[1] that the raw unconceptualized data of sensory experience constitute the 'immediately given', as yet untinged by the activities of the mind and appearing before our consciousness independently of what we think. Epistemological problems typically arise only over the subsequent interpretation, conceptual fixation, abstraction and generalization with respect to these sensory data, that is, over the question of just how concepts relate to percepts and percepts to sense-data, and just how the results of *conscious* processing relate to the immediate data of consciousness. However, the basic assumption is unwarranted. It presupposes a Cartesian theory of mind according to which the mind—no matter how described or how referred to—is necessarily conscious of all its activities. This theory is by no means self-evident. Its denial does not involve inconsistency except when the mind is arbitrarily *defined* in Cartesian terms.[2] Thus there is no guarantee that the sense-data which appear to be immediately given are not in fact mediated by subliminal operations of the mind

---

1 The attempt to reduce all empirical knowledge to 'pure' sense data (providing a non-inferential base) plus logic (and set theory) was characteristic of logical atomism and of Carnap's *Aufbau*. To realize the ideal embodied in this program represented something like the Holy Grail of analytic epistemology ardently, but vainly, sought after by an entire generation of twentieth century philosophers. Cf. B.A.W. Russell, *Our Knowledge of the External World*, (London 1914); M. Schlick, *Allgemeine Erkenntnislehre*, (Berlin, 1918); L. Wittgenstein, *Tractatus Logico-Philosophicus*, (London, 1922); R. Carnap, *Der logische Aufbau der Welt*, (Berlin, 1928); C.I. Lewis, *Mind and the World Order*, (New York, 1929); H.H. Price, *Perception*, (London, 1932); A.J. Ayer, *The Foundations of Empirical Knowledge*, (London, 1940); B.A.W. Russell, *Inquiry into Meaning and Truth*, (London 1940); C.I. Lewis, *An Analysis of Knowledge and Valuation*, (La Salle, 1946); H. Reichenbach, *Experience and Prediction*, (Chicago, 1947); B.A.W. Russell, *Human Knowledge: Its Scope and Limits*, (London, 1948).

It is not difficult to see that this program of twentieth century logical empiricism has a 'Cartesian' motivation and is in fact epistemologically continuous with the Fregean program for arithmetic and the Russellian program for all of mathematics (with Leibniz as *auctor intellectualis*). For Frege's and Russell's logicism similarly aspired to reduce the foundations of mathematics to pure logic (plus set theory) in order to render (in this case) mathematics clean, pure and certain.

2 Descartes held that the essence of a mind is consciousness, or to be conscious. He reached this thesis by consciously extending the received French and Latin usages of the words *'pensée'* and *'cogitare'*, respectively, and of their various cognates so as to cover all the operations of will, intellect, imagination and of the senses [cf. HR II 52; also cf. Anthony Kenny, *Descartes: A Study of His Philosophy*, (New York, 1968), pp. 68-9]. Now since to think is to be conscious, all mental operations or 'thoughts' must be conscious operations as well.

interpreting the sensory excitations even before we are conscious of perceiving anything.

Helmholtz's controversial theory of unconscious inferences embodies such a non-Cartesian analysis. However, once the Cartesian premise has been abandoned, the futility of strictly formal or strictly transcendental approaches in epistemology and the indispensability of a genetic approach follow almost immediately. A phenomenalistic analysis of perception finds its rock bottom in the simple sense-data of sensory experience which we are immediately conscious of and which cannot be further analyzed. Such introspectively accessible sense-data may be expressed in primitive statements which can be regarded as the basis for further conscious cognitive processing. However, if it is hypothesized that these data themselves are unconsciously obtained by inference rather than by direct assimilation of a given sensory input, then only physiological investigation or experimental psychology, not a priori analysis, can identify the particular sensory excitation that provides the middle term of the habitual inference, whose conclusion alone expresses a conscious experience. Therefore, since complicated and largely involuntary, though experientially conditioned, processes of interpretation precede most of our so-called immediate experiences, we would be mistaken in regarding the latter as the basic units of empirical knowledge. Indeed, granted the operation of such 'proto-intellectual' activities in human perception, we may more readily contemplate the idea that similar subliminal processes may constitute a substantial part of our entire intellectual life. The very notion of such 'ratiomorphic' foundations of man's celebrated logical faculties may fly in the face of 'Reason' as the formalists would have it and be dismissed by them as mere psychologism, yet the assertion of its absurdity would seem to require more persuasive arguments than the mere observation that its acceptance would render the formalists' a priori analyses and their justificationist objectives hopelessly inadequate.

Contrary to Helmholtz's interpretation of it, Kant's philosophical enterprise was meant to be strictly transcendental. It did not aim at identifying any factual conditions of the perceptual and cognitive apparatus. Rather, it specified those conditions that were claimed to be necessary for any experiential knowledge of reality at all. If those conditions did not obtain, there could be no *experience* to falsify the results of Kant's analysis. Thus Kant investigated those a priori sources of human cognition that are *logically*, and not just psychologically or physiologically, prior to whatever material knowledge man may obtain.

> All knowledge which is not so much concerned with objects, but rather with the nature of our knowledge of objects insofar as this is possible a priori, I call transcendental knowledge.[1]

By its very nature, therefore, transcendental knowledge would be invalidated if it rested essentially on empirical considerations.

In the mid-nineteenth century the predominant trend was to interpret Kant psychologically rather than logically. The essence of a priori knowledge was sought in its being conditioned by contingent psychological and physiological structures. There is indeed sufficient ambiguity in Kant to lend partial support to this view. Helmholtz certainly shared this interpretation, as had his teacher in physiology Johannes Müller before him.[2] This encouraged an unmistakable tendency in Helmholtz's thought to naturalize Kant's rationalist ideas, particularly his conception of the aprioricity of space and, to a controversial degree, his theory of causality.

2. Müller's Principle of Specific Sense-Energies.

One of the most influential physiological theories that stimulated much of Helmholtz's empirical research and epistemological theorizing was the doctrine of specific sense-energies as developed by his teacher in physiology at the Royal Friedrich-Wilhelm Institute for Medicine and Surgery in Berlin, Johannes Müller. Scattered throughout his writings we find references to this principle which—though admittedly not without vaguer forerunners and suggestive philosophical hunches pointing in the same direction—was regarded by Helmholtz as firmly established only by modern physiology.

The question as to how the animate body differentiates and coordinates information received through the different senses to give an integrated single view of the world as well as suitable behavior in relation to it was of course a classical problem known to antiquity. Regarding the question of

---

1 "Ich nenne alle Erkenntnis transzendental, die sich nicht sowohl mit Gegenständen, sondern mit unserer Erkenntnisart von Gegenständen, insofern diese a priori möglich sein soll, beschäftigt." [Immanuel Kant, *Kritik der reinen Vernunft*, (Darmstadt, 1968), I, Einleitung, VII, p. 63].

2 Not surprisingly, Rokitansky nicknamed Kant "der Mann der Physiologen"; cf. A. Riehl (1924), p. 225.

differentiation the atomists[1] emphasized specific irritability rather than specific energy. This tendency continued among traditional empiricists in general. According to the former principle the senses report different information because they are so constructed as to be selectively sensitive to different kinds of stimuli or 'images' from the surrounding world. According to the latter (Stoic) principle developed by Galen sensory differentiation resulted from differential internal responses of the specific *pneuma* sent by the soul to the special sense organs to any form of stimulation.[2]

Descartes, too, adopted the principle of specific sense energies expressed in Galen's theory of the specific *pneuma*. Thus

> people hit in the eye think they see a great number of fiery flashes in front of them, in spite of shutting their eyes or being in a dark place.... The same force might cause one to hear a sound, if applied to the ears, or to feel pain, if applied to other parts of the body.[3]

Indeed, Descartes took this as a confirmation of his view that the objective nature of light and color "simply consisted in various sorts of motion"[4] brought about by varying states of mechanical pressure and producing different sensations according to the different motions generated in a particular nerve. Thus the same sensation could be produced by different kinds of external or even internal stimuli as long as they caused a similarly patterned motion on the walls of the pineal gland. Taking this view to its logical (and skeptical!) extreme Mersenne even argued that angels could make us see and hear 'external' objects by suitable stimulation of our (internal) nerves.[5]

---

1 Cf. J.I. Beare, *Greek Theories of Elementary Cognition From Alcmaeon to Aristotle*, (Oxford, 1906), pp. 25-37.

2 Galen, *De placitis Hippocratis et Platonis*, VII, pp. 4-5.

3 *Dioptrics*, VI, AG 247.

4 Ibid.

5 F. Marin Mersenne, *Harmonie universelle*, (Paris, 1636-37), I, Prop. 52. In fact, a more radical conclusion could be drawn. When in the course of the 19th century Müller's theory had developed into a 'projection' theory of sensory qualities (according to which the receptor fields of the sense organs are projected upon the central nervous system in the sense that the afferent fibers lead to different parts of that system), E. Duboys-Reymond imagined that if only one could cross-connect, say, the auditory and the optic nerves, we would be able to hear lightning and to see thunder. Hering took this to be an "absurd consequence" of the view that different effects of nervous activity are adequately explained in terms of differential terminal parts or that functional differentiation

However, it was not until the 19th century that important *physiological* details and arguments of a less speculative nature would be brought to bear on the still unsettled philosophical dispute. The Bell-Magendie Law, in particular, provided an effective stimulus for physiological research into the area of sensory perception.[1] The dichotomy of nervous activity into sensory and motor activity, as asserted by this law, encouraged physiologists to think that the mind's sensations were as much their business as the muscles' movements.[2] Moreover, in the introductory chapters to his essay of 1811 Bell already clearly anticipated Müller's later doctrine.[3]

Finally, then, Johannes Müller formulated his, now classical, statement of the theory, promulgating what he coined the principle of specific "energies" of nerves, in which he gave comprehensive physiological meaning to the old philosophical doctrine of sensory specification. Thus he paved the way for the mid-19th century synthesis of sense physiology on the one hand and philosophical psychology and epistemology on the other.

> V. Sensation consists in the 'sensorium' receiving through the medium of the nerves, and as the result of the action of an external cause, a knowledge of certain qualities or conditions, not of external bodies, but of the nerves of sense themselves; and these qualities of the nerves of sense are in all different, the nerve of each sense having its own peculiar quality or energy....
>
> VIII. The immediate objects of the perception of our senses are merely particular states induced in the nerves, and felt as sensations either by the nerves themselves or by the sensorium; but inasmuch as the nerves of the senses are material bodies, and therefore participate in the properties of

---

exclusively resides in either the central or the peripheral apparatus. [cf. Ewald Hering, 'The Theory of Nerve Activity', (1898), in id., *On Memory*, (Chicago, 1913)].

1 Charles Bell was the first to discover that the motor and the sensory functions of the nervous system are actually separated, the motor nerves leaving the spinal cord by the anterior roots. He reported his discovery in his *Idea of a New Anatomy of the Brain*, (London, 1811). François Magendie reached the same conclusion in 1822 as a result of more convincing experiments (cf. his 'Expériences sur les fonctions des racines des nerfs rachidiens', *Journal de physiologie expérimentale et pathologique*, (1822), 2, pp. 276-9; and his 'Expériences sur les fonctions des racines des nerfs qui naissent de la moëlle épinière', ibid., pp. 366-71).

2 Cf. Edwin G. Boring, *Sensation and Perception in the History of Experimental Psychology*, (New York, 1942), p. 8.

3 Charles Bell (1811); also in HB 23 ff.

> matter generally occupying space...they make known to the sensorium, by virtue of the changes thus produced in them by external causes, not merely their own condition, but also properties and changes of condition of external bodies. The information thus obtained by the senses concerning external nature, varies in each sense, having a relation to the qualities or energies of the nerve.[1]

However, the "insidious and insistent belief"[2] that experience and stimulus are necessarily alike, a view which two centuries of progressive philosophical thought since Descartes had in vain sought to dispel, persisted in part even in Müller's classical statement. Thus Müller also recognized "specific irritability" for certain influences:

> The nerves of the senses have assuredly a specific irritability for certain influences; for many stimuli, which exert a violent action upon one organ, have little or no effect upon another: for example, light, or vibrations so infinitely rapid as those of light, act only on the nerves of vision and common sensation; slower vibrations, on the nerves of hearing and common sensation, but not upon those of vision; odorous substances only upon the olfactory nerves.[3]

However, when he added that "the external stimuli must therefore be adapted to the organ of sense—must be 'homogeneous'"[4] he came again dangerously close to the naïve traditional image theory of perception which his very own theory had been designed to supersede.

Now Helmholtz, while recognizing the history of Müller's principle and giving due regard to similar contemporary theories developed by Young[5] and Bell[6], esteemed especially Müller's contribution to physiology as almost equally significant as that of Newton's law of gravitation to mechanics—a mildly unsettling claim, I must confess, which is the more conspicuous as Helmholtz does not ordinarily indulge in hyperboles.

> Müller formulated with classical perfection what people until then had only surmised from the data of daily experience and

---

1 Johannes Müller, *Handbuch der Physiologie des Menschen*, (Coblenz, 1838), V, 'Introduction'; tr. by William Baly as *Elements of Physiology*, (London, 1842), II, 'Preliminary Considerations', pp. 1065 ff.

2 M. Wertheimer (1922); quoted in E.G. Boring (1942), p. 5.

3 Loc. cit.

4 Ibid.

5 Thomas Young, 'On the Theory of Light and Colours', *Philosophical Transactions of the Royal Society of London*, (1802), 92, pp. 18-21.

6 C. Bell (1811).

> had tried to express vaguely, mixing truth with falsehood, or what had been clearly formulated only for special, more narrow areas, as Young had done for the theory of colors and Bell for the motor nerves. I am inclined to value this [Müller's] scientific achievement as highly as the discovery of the law of gravitation.[1]

In short, Müller's principle asserts that any given nerve has specific effects independent of the nature of its stimulation. Whether a given stimulation will yield a muscular contraction, a secretion or a sensation is determined solely by the neuro-physiological nature of the nerve (or nerves) affected by it. Motor, glandular or sensory nerves, if excited, will invariably produce the same specific results. The situation regarding the nerves of the five senses is no different. Each one of them has a 'specific energy' which may vary in intensity but not in kind. The optic nerve, whether excited by pressure or by radiation, by electric impulses or by strain, invariably produces the sensation of light. In view of the variety of circumstances under which we may have luminous sensations physiologists for a long time had believed that the eye was capable of producing real light. Müller, however, showed that all sensation of light has a single proximal cause, the activation of the *opticus*.

Conversely, the same objective stimulus gives rise to altogether different sensations as it stimulates different sense-nerves. What the eye perceives as light, the skin may perceive as heat; what the skin may feel as vibration, the ear hears as tone. It was a long time before physicists were convinced that the phenomena of light and heat, which we sense as so radically different as to defy comparison, are in fact essentially similar and partly identical physical processes. If you like a paradoxical formulation: light becomes light only when it strikes a live retina; otherwise it is mere vibration of the ether.[2] The subjectivity of our sense impressions is further evidenced by the fact that our sense nerves respond only to a small segment of the total range of values of any given measurable impulse. The same physical events, therefore, cease to produce 'standard' sensations whenever their measurable properties exceed a certain threshold value relative to a given sensory system.

---

1 "Was man bis dahin aus den Daten der täglichen Erfahrung geahnt und in unbestimmter, das Wahre mit Falschem vermischender Weise auszusprechen gesucht, oder nur erst für einzelne engere Gebiete, wie Young für die Farbentheorie, Bell für die motorischen Nerven fest formuliert hatte, das ging aus Müller's Händen in der Form klassischer Vollendung hervor, eine wissenschaftliche Errungenschaft, deren Wert ich der Entdeckung des Gravitationsgesetzes gleichzustellen geneigt bin." (VR II 181-2).

2 VR I 98.

Müller's principle raised the problem of just how many distinct senses there are. This Helmholtz solved by means of the criterion of continuous variation. Any two primitive sensations can be categorized according to whether or not the distinction between them is commensurable. In Helmholtz's terminology, incommensurable sensations (like blue and bitter) differ in *mode*, while sensations that can be compared ordinally (like blue and ultramarine blue) differ only in *quality*. The sensations connected by qualitative continua belong to the same modality. Colors lie in a single modality because they can be placed in a single 3-dimensional continuum, the color solid. Tones form a modality, too. But touch does not, for at least pressure, temperature and pain are discrete.[1]

Sensations of the same modality constitute a *quality-sphere*. Physiologically they belong to the same sense-organ. But beyond that we cannot assert anything. That is, we cannot single out any specific set of physical events as the set of external causes of the sensations that belong to a given quality-sphere. As a consequence of Müller's principle of specific sense-energies there exists no standard correspondence between the modality of a given sensation and the nature of its stimulus. Thus a profound epistemological problem arises. Perceptual knowledge of the external world, if possible at all, will have to be based on *equivocal* sign/significate relationships. The real magnitude of this problem was not immediately obvious to all concerned. Helmholtz was one of the first philosopher-scientists to fathom the profundity of the implied epistemological problem and to contribute effectively to its solution.

Müller's principle pertained only to the mutual relations of different sense-nerves. Inspired in part by Young's theory of colors Helmholtz extended the applicability of the principle to the relations of different nerve fibers of one and the same sense-organ. Thus Müller's doctrine generated the basic ideas underlying Helmholtz's physiological optics and physiological acoustics, as well as his epistemological theory according to which sensations constitute a system of empty symbols which do not copy objective reality and whose meanings therefore must be *learned*.[2] Let me first briefly summarize the results of Helmholtz's further empirical research in the following two sections and then proceed to discuss his epistemological elaborations.

---

1 The subsequent problem of determining the number of separate elements of sensation proved more controversial. Fechnerian psychophysics had furnished the 'just-noticeable-difference' as a means for fractionating qualitative continua and thus of counting sensations. Lotze and Stumpf, however, argued that sensation can be regarded as a continuous function of the stimulus, and that the j.n.d. was an artifact of observation [cf. E.G. Boring (1942), pp. 10, 38].

2 PO III 533; VR I 392 f.; VR I 317 ff.

3. Helmholtz's Theory of Color Vision.

The theory of color vision developed by Helmholtz over a century ago has remained the dominant one up till fairly recently, and is still largely consistent with ingenious recent amplifications and refinements especially in the area of the theory of color constancy.[1] While doing research on complementary colors and color mixtures Helmholtz rediscovered Young's hypothesis, which, as he said, had been "buried" in the Transactions of the Royal Society and forgotten for decades.[2] The problem facing Young was to explain how the overwhelming complexity and diversity of physical light is being 'translated' into the astounding simplicity of color vision. It was towards the solution of this question that he had taken a decisive step. He boldly conjectured that there must be three kinds of retinal nerve fiber corresponding to the basic colors red, green and violet. When light of any frequency within the visible spectrum strikes the retina, all these fibers are affected by it, but they are selectively activated, the intensity of their respective responses varying with the frequency:

> for instance, the undulations of green light...will affect equally the particles in unison with yellow and blue, and produce the same effect as a light composed of those two species.[3]

Helmholtz immediately recognized Young's hypothesis as interpretable in terms of Müller's principle. Before Young, physiologists had simply assumed that the optic nerve was capable of producing very different luminous sensations without inquiring into the reason why precisely *that* system of color sensations emerges as is produced by the normal eye.

Natural sources of light and reflecting surfaces rarely emit pure light of constant frequency. Most natural light is compound. It can be analyzed into numerous waves of continuously varying frequencies. Since the properties of such light depend on the relative amplitudes of all these separate wavelengths, its physical quality in general is representable only as a function of indefinitely many variables. It is absurd to suppose that there are retinal

---

1 Land's so-called retinex theory of color vision, according to which color is determined by three lightnesses, each computed from comparisons using intensity information from the entire image, bears all the marks of a definitive treatment; cf. E.H. Land, 'The Retinex', *Am. Scientist*, vol. 52 (1964), pp. 247-64; id., 'The Retinex Theory of Color Vision', *Sci. Am.*, Vol. 237 (1977), 6, pp. 108-28.

2 VR I 312.

3 Thomas Young (1802), HB 13. For the (traditional) error in the details of the theory cf. p. 139, n. 2. Helmholtz criticizes Young's original choice of red, yellow and blue as fundamental colors in OP II (1911), p. 118.

points corresponding to each of the innumerable frequencies of light-waves within the visible spectrum. On the other hand it is hardly more plausible to hypothesize that each single terminal fiber of the optic nerve, when struck by a ray of natural light, is capable of serving as a receptor-organ for all the numerous light waves of different frequencies of which that ray of light is composed, and of simultaneously transmitting multiple signals representing their relative intensities.[1]

Thus the factors determining the objective quality of most natural light are usually extremely complex. Yet the variables responsible for the subjective nature of the sensation of light as color may be fairly limited in number. Experiments with pigmental mixtures, known of course since ancient times, suggested the idea of three so-called fundamental colors. When mixed in varying amounts these colors turn out to produce almost all other colors within the spectrum. Thus, all our color impressions may perhaps be regarded as functions of only three independent variables.[2]

It is important to bear in mind, however, that these basic colors have no objective significance.

> A reduction of the colors to three fundamental colors can never have more than merely subjective significance, it can involve no more than a reduction of the *color sensations* to three *fundamental sensations*.[3]

---

1 Cf. Thomas Young (1802); also cf. HB 13.

2 Mixtures of prismatic colors (i.e., compounds of pure homogeneous light of different frequencies) of course do not yield the same results as pigmental mixtures, as was traditionally assumed (up to Helmholtz's time!). The most striking discrepancy is the fact that painters mix blue and yellow to get green while a mixture of the corresponding prismatic colors yields pure white as the colors involved are complementary. It was Helmholtz himself who first discovered this discrepancy and formulated the (ingenious) theoretical explanation of this phenomenon. Cf. Hermann von Helmholtz, 'Ueber die Theorie der zusammengesetzten Farben' (1852), WA II, pp. 15 ff.

3 "Eine Reduktion der Farben auf drei Grundfarben kann immer nur subjektive Bedeutung haben, es kann sich nur darum handeln, die *Farbenempfindungen* auf drei *Grundempfindungen* zurückzuführen." [PO II (1911); (1962) 142-3]. This, now commonsensical, notion still needed articulation and explanation some hundred years ago. Thus Brewster sought the explanation of the well-known fact that there are three fundamental colors not in the constitution of man (as Young had been the first to suggest), but in the nature of light. He maintained that for every wavelength there were three different kinds of light, mixed merely in different proportions so as to give the different colors of the spectrum. For Helmholtz's refutation cf. his 'Über Herr D.Brewsters neue Analyse des Sonnenlichts', *Pogg. Ann.*, (1851), LXXXVI, p. 501. Also cf. F. Bernard, 'Thèse sur

Young's conjecture, then, provided an elegant account of the 'simplicity' of color vision. Moreover, Helmholtz was particularly struck by the fact that Young's hypothesis was clearly interpretable as a special application of the principle of specific nerve energies:

> Just as tactile sensation and visual sensation of the eye demonstrably belong to different nerve fibers, Young here assumes the same for the sensation of the various fundamental colors.[1]

Just as the nerves of different senses respond uniformly according to their proper mode of sensation to a vast array of objectively different stimuli, so does each retinal fiber respond with varying intensity to light waves of any frequency within the visible spectrum while the specific color sensation it yields is invariably the same in accordance with its 'specific energy'.

Helmholtz also followed Young in ascribing no other properties to the optic nerve than those of the motor nerves, viz., the contrast between a state of rest and a state of activity.

> The dark sensation corresponds to the state of rest of the optic nerve, that of colored or white light corresponds to an excited state thereof.[2]

---

l'absorption de la lumière par les milieux non crystallisés', *Ann. de chim.*, (3) (1852), XXXV, pp. 385-438.

1 "Wie Tastempfindung und Gesichtsempfindung des Auges nachweislich verschiedenen Nervenfasern zukommt, wird hier dasselbe auch für die Empfindung der verschiedenen Grundfarben angenommen." [PO II (1911), p. 121; cf. VR I 313].

2 "Die Empfindung von Dunkel entspricht dem Ruhezustand des Sehnerven, die von farbigem oder weissem Licht einer Erregung desselben." [PO (1885-1894) II 345].

Hering, in his 'Theory of Nerve Activity', (1898; Chicago, 1913) took strong exception to the implied view which, with the combined authority of Helmholtz, DuBoys-Reymond and Donders, had become virtually 'unassailable doctrine', viz., that all nervous activity is essentially homogeneous. Instead he assumed that nervous substances are capable of multiple functions and specific differentiation. This assumption was a basic premise of Hering's own theory of color vision formulated in 1874 [E. Hering, *Zur Lehre vom Lichtsinn*, V ff.; *Grundzüge einer Theorie des Lichtsinns*, (1874-75); *Sitzungsberichte der Wiener Akademie*, mathem.-naturw. Kl. LXIX, (1874), p. 131]. This theory identified six specific energies arising from three nervous substances each capable of two antagonistic and mutually incompatible processes associated with the complementary color sensations: the red/green substance, the yellow/blue substance and the white/black substance.

Various attempts at a 'higher synthesis' of the two opposing theories of color vision were made, e.g. by Donders (1881) who, in following out Mach's psychophysical parallelism, suggested that the Young-Helmholtz theory might have merely retinal application while the tetrachromatic theory of the Hering school might refer to an indepen-

Thus Helmholtz was able to show that in order to account for the phenomena of apparent coloration of negative after-images all one needed was a special amplification of the theory of retinal fatigue which Fechner had proposed to explain the rise of these after-images themselves. For example, when we focus our eyes sufficiently long on a blue-green object which is placed on a red ground and then suddenly removed, its after-image will have a red color more saturated than the red of the ground. The exposure to the blue-green color has predominantly stimulated, and thereby fatigued, those fibers that are sensitive to green and violet. The affected part of the retina is thus made less sensitive to whatever non-red components the light of the ground may contain as well as to the weak stimulating effects which even homogeneous red light ordinarily exerts on the non-red-sensitive fibers of the optic nerve.

The scientific merits and demerits of Helmholtz's theory of colors cannot be extensively discussed here. But his theory has not remained unchallenged.[1] Though it provided a satisfactory explanation of quite a few optical phenomena about which empirical data were available at the time, especially in the areas of complementary colors and color mixtures, his theory became increasingly cumbersome as it was invoked to account for such well-investigated phenomena as partial or total color blindness.

4. Helmholtz's Theory of Physiological Acoustics.

Helmholtz also employed the principle of specific nerve energies in his theory of physiological acoustics. His research built upon work done in the first half of the nineteenth century by Fourier in mathematics and by Ohm in physics. Fourier's principle established that certain continuous periodic functions or curves are representable as summations of series of sine curves or uniform waves which vary in amplitude, phase relation and wavelength,

---

dent cortical mechanism. Von Kries developed this suggestion into what he called a 'zonal theory' which similarly supposed "that the sensations of vision may be aroused by two different mechanisms more or less independent of each other" (PO (1962) II, Appendix, p. 432). Finally, a highly speculative development providing an attractive break-through of the theoretical stalemate between the 'trichromatists' and the 'tetrachromatists' was given by Christine Ladd-Franklin who gave an evolutionary twist to the opposing theories by assuming a 3-stage development of the color sense (*Zft. f. Psychologie*, (1892), Bd. 6, p. 4).

1 Cf. preceding footnote.

the wavelength of the members of the series being even fractions (1/1, 1/2, 1/3...) of the length of the wave being represented. In the case of sound this corresponds to the fundamental note and its several harmonics. Ohm then showed that the mathematical analysis, established by Fourier's principle, of certain complex curves (which can be used to represent sound waves) into a series of simple sine curves or uniform waves can also be performed physically by using various kinds of resonators. He went on to give a psycho-physiological interpretation of his acoustical law. Perhaps, he suggested, the ear is just such a resonating mechanism by virtue of which it analyses compound notes or sounds into their simple harmonic components.

Helmholtz picked up this suggestion. At first (1857) he located this mechanism in the microscopic rods of Corti (which had been discovered recently) and treated these as a graded series of resonators. Later (1869) he chose for the resonating mechanism in question a membrane in the cochlea consisting of a series of fibers of graded length. Again, Müller's law proved to be a surprisingly effective explanatory tool. For Helmholtz's physiological acoustics is in essence a resonance theory coupled with a special and novel application of Müller's principle. Whereas the eye had been hypothesized to produce only three different kinds of nervous energies, Helmholtz now postulated that the ear contained over 5000 slightly different kinds of receptor-organs, its analytic power with respect to complex sound waves consequently being incomparably greater than the eye's power to analyze compound light. Thus, in the sensation of complex sounds, the ear preserves the results of its analysis with a high degree of reliability: two chords made up of different musical notes will never appear identical to the ear whereas to the eye different aggregates of compound colors may look exactly alike.[1]

5. The Philosophical Significance of the Principle of Specific Sense-Energies.

In the preceding sections I have summarized some of the results of Helmholtz's physiological investigations. If these results have left many epistemological problems unsolved, at least they have served to bring them into sharper focus. The essential interdependence of, and the indispensability of interaction between, the philosophical and the empirical parts of the

1 Cf. PO II (1962), p. 120.

epistemological enterprise Helmholtz always presupposed as an almost unassailable doctrine from the very beginning. Thus he writes in 1885:

> The point at which philosophy and science are most intimately intertwined is the theory of sensory perception in man. Therefore I will try to present to you the scientific results for that sensory organ whose operations have so far been investigated in the most complete fashion possible; that organ is the eye. You can then judge for yourself how the empirical results relate to those of philosophy.[1]

The extensive applicability of Müller's principle as elaborated in Helmholtz's physiological optics and physiological acoustics had provided him with his most fundamental epistemological insight and with the point of departure for his philosophical reflections. Constituting the backbone of his so-called empirical (more properly: cognitive) theory of perception this basic thesis states that the very material in the afferent nervous system for the subsequent formation of perceptions of reality does not copy external reality but constitutes a *naturally given system of symbols* which the human organism must *learn* to interpret for the development of adaptive patterns of behavior.

Helmholtz's scientific and empiricist turn of mind and his tendency to interpret Kant's transcendental theories of the a priori in terms of organic and psychological structures of data-processing made him regard the above result as an empirical extension and corroboration of Kant's views on the constructive character of human experience. The implied relations, however, are far from obvious. The assumption of specific sense-energies suffices to explain—indeed very elegantly—the subjectivity of the sense-qualities but it is certainly not a necessary condition of the latter. This assumption is no more than a useful empirical hypothesis which must be established on independent grounds before it can be said to lend support to a subjectivist philosophical view of a corresponding kind. However, since the truth of Müller's principle is by no means necessarily implied by any such subjectivist theory—and certainly not by Kant's—it might be refuted without its falsehood affecting to the slightest degree the validity of the philosophical theories under consideration. On the other hand, its truth of course can have no bearing

---

1 " Der Punkt an dem sich Philosophie und Naturwissenschaften am nächsten berühren ist die Lehre von den sinnlichen Wahrnehmungen des Menschen. Ich will mich daher bemühen, Ihnen die Resultate der Naturwissenschaften für das Sinnesorgan darzulegen, dessen Verrichtungen bisher am vollständigsten untersucht werden könnten; dieses Organ ist das Auge. Sie werden dann selbst urteilen können in welchem Verhältnisse hier die Ergebnisse der Erfahrung zu denen der Philosophen stehen." (VR I 90).

whatsoever on the plausibility of these philosophical theories just in virtue of their having the same consequence. In order for Müller's and Helmholtz's empirical theories to have some corroborating relevance with regard to Kant's philosophy, therefore, one is required to show not merely their probability or acceptability—assuming this can be done satisfactorily—but also their being somehow a faithful translation in physiological terms of the general philosophical ideas which led Kant to assert the subjectivity of all human experience. This they must have tacitly assumed. But is their assumption historically correct? I think not.

Kant's views were obtained by transcendental analysis and proof. Even if the qualities of sensation (which for him constituted the real *content* of our experience) were objective—or, at any rate, stood in some unique correspondence to the various surface characteristics of an independent reality—still their informational content could become effective only *after* they had been interpreted as the qualities of physical objects in space and time. But, according to Kant, space and time as such are certainly not objects of immediate experience. It is not from experience that we have derived these concepts. Rather, the very idea of a perceivable quality without any spatial and temporal attributes involves a *logical* contradiction.

Thus the suggestion implied by Müller and Helmholtz that the specific sense-energies of the various afferent nerve systems of different modes constitute, in effect, an enlarged set of a priori forms of perception of the sort envisaged by Kant is plainly false.[1] In Kant's view, the a priori forms of perception

---

1 Thus Helmholtz wrote:
"Kant achieved the most extraordinary progress for philosophy in that he searched for, and identified, the law under consideration as well as the other innate forms of perception and laws of thought. As I have already mentioned before, he thus achieved for the theory of representation in general the same as what physiology due to Johannes Müller had achieved by empirical means in a smaller area for the immediate sensory perceptions."
("Es war die ausserordentlichste Fortschritt, den die Philosophie durch Kant gemacht hat, dass er das angeführte Gesetz und die übrigen eingeborenen Formen der Anschauung und Gesetze des Denkens aufsuchte und als solche nachwies. Damit leistete er, wie ich schon vorher erwähnte, dasselbe für die Lehre von den Vorstellungen überhaupt, was in einem engeren Kreise für die unmittelbaren sinnlichen Wahrnehmungen auf empirischen Wegen die Physiologie durch Johannes Müller geleistet hat." ['Ueber das Sehen des Menschen', (1855), VR I 116].

cannot simply be reduced to any psychological or physiological substratum. They are logically necessary conditions of any perception as such and—in virtue of their synthetic functions—sources of synthetic a priori propositions. By contrast, the principle of specific sense-energies and its various applications, even if universally valid, would still not be necessary nor—in spite of its synthetic functions—would it give rise to propositions that are valid a priori. It is not a necessary condition for the occurrence of any experience at all that it be organized in accordance with Müller's law. This law merely spells out the general physiological conditions governing the sensory transmission of physical information in man. It thereby exposes the limitations of our perceptual apparatus and it sheds light on the irreducible indeterminacy inherent in correlating given physiological and psychological stimuli: the qualities of our sensations—i.e., the *contents* of our experience and not just the *form* under which, according to Kant, they must necessarily be perceived—are not in one-to-one correspondence with the qualities of antecedent physical events but depend to a large extent on the neurophysiological properties of our senses.[1] Thus the subjectivist view of human experience which may be said to be empirically corroborated by Müller's principle, if true, is not at all the characteristically Kantian one but rather one that generally follows from a *mechanical* conception of nature: the kind of effects wrought by any given cause depends on the typical response-capacities of the physical system on which that cause operates.

Throughout the history of philosophy, from the disintegration of early materialism to the debates surrounding modern reductionism, we find a virtually inevitable link between strictly mechanical world views and the tendency to emphasize, with increasingly idealist overtones, the subjective components of sense-experience. Thus, by the force of its own presuppositions it seems to be the historical fate of philosophical materialism to ultimately collapse into pure sensationalism. What unites the two schools of thought is their common dedication to atomist models of explanation. Thus, in the latter case the whole of consciousness is referred back to the sensations just as in the former case the whole of nature is dissolved into atoms and their motions. At the bottom of their divergence we find of course the perennial mind-body problem. And in the ensuing struggle materialism is at a considerable dis-

---

1 For an early statement on perceptual subjectivism and, despite of it, on our continued epistemic access to the structural properties of the real world, cf. Helmholtz, 'Ueber die Natur der menschlichen Sinnesempfindungen' (1852), in WA II 608-9. Here Helmholtz already expresses the doctrine that sensations are only symbols for real external states of affairs which we, in the capacity of sentient organisms as well as in the capacity of measuring physicists, can only reconstruct and approximate in mediate ways, but never immediately.

advantage from the outset. For the problem it faces of how to explain the emergence of conscious sensations from mere material motion appears to be all but insurmountable. As Lange observes, materialism "can hardly close the circle of its system without borrowing from idealism".[1] By contrast, the phenomenalist does not face that systematic difficulty. Without leaving the confines of his own territory, i.e., the intelligible rather than the 'real', self-existent world, he can consistently account for our ideas of matter and force, of atoms and their motions as resulting from our intellectual dispositions towards the primordial sensations. The astronomer Zöllner thus stressed the fact that the sensations are the material out of which the world of external things constructs itself.

> ...the phenomenon of sensation is a much more fundamental fact of observation than the motion of matter, which we are obliged to attribute to it as the most universal quality and condition of the intelligibleness of sensuous changes.[2]

Even Einstein, in his earlier 'empiriocriticist' period, reduced the whole of experimental physics to the observation of point-coincidences through pointer readings. Thus, the key concept in the early part of his 1905 paper is introduced with unmistakably sensationalist overtones. Einstein wrote:

> ...all our judgments in which time plays a part are always judgments of *simultaneous events*. If for instance I say, 'that train arrived here at seven o'clock', I mean something like this: 'The pointing of the small hand of my watch to seven and the arrival of the train are simultaneous events'.[3]

Unassailable though the sensationalist position may *prima facie* seem to be, it does, however, present difficulties of its own. Once the senses have been denied glimpses into the 'real' world, the inward gaze tends to become obsessive. As Russell had it, what the physiologist sees when he is examining someone else's brain is part of his own brain, not part of the brain he is examining. Such bizarre consequences may very well be taken as reducing sensationalism to absurdity.[4]

---

1 Friedrich Albert Lange, *The History of Materialism*, (1865), E.C. Thomas (tr.), with an introduction by B. Russell, (London, 1925).

2 Quoted in ibid., II, p. 326.

3 Albert Einstein, 'Zur Elektrodynamik bewegter Körper', *Ann. d. Physik*, 17 (1905), p. 893. For a stimulating account of Mach's (early) influence of Einstein and the gradual emergence of Einstein's explicit antipositivism, cf. Gerald Holton, 'Mach, Einstein, and the Search for Reality', in id., *Thematic Origins of Scientific Thought*, (Cambridge, 1973), pp. 219-59.

4 Russell's 'absurd' phenomenalism, however, often seems somewhat more justified especially in the context of inquiry into the physiology of the senses. Thus, in PO II, p.

Thus, in their conception of the universal reign of law governing all changes, materialism and physical reductionism alike seem to contain the very seeds of their ultimate philosophical disintegration. For as the effect has to be an event different from the cause, the sensation—whether or not a mere brain state, yet undeniably our only access to the external world—embodies information necessarily different from whatever brought it about. It is here, essentially, in a philosophical climate that sponsors mechanical conceptions or a search for last principles governing the ever-changing relations among the ultimate constituents of the unitary universe of events, that the systemic cause is to be found for the *prima facie* puzzling transition—so amply illustrated by the history of ideas—from austere systems of *materialism*, which demand that a satisfactory explanation subsume the laws governing all motion; via *sensationalism*, which demands that the intelligible world be reconstructed solely from indubitable sense-data as basic units; to *subjective idealism* and *solipsism*, in which all that is left of 'reality' is a single subject and his phenomenal world. Usually the sequence finds its ultimate stage when the pained epistemologist turns vice into virtue by espousing some form of *skeptical relativism*, which professes blissful ignorance and trades the ever receding hope of hidden truth and meaning for the dubious delight of aimlessly splashing around in an uncharted sea without much more to care for than to keep floating and to derive incessant joy from it.

Locke tried to hold untenable ground between Hobbes and Berkeley. Helmholtz, weary of metaphysical speculation and steeled by a scientific determination to remain as close to the data of his research as possible, similarly steers a delicate course in his symbol-theory of sensations between mechanical materialism on the one hand, and on the other hand a Neo-Kantian idealism which declares the noumenal reality of Kant's *Ding-an-sich* to be pure transcendental illusion. The apparent ambiguities of this position ignited the scorn of Lenin who observed that Helmholtz's "'half-hearted materialism' with Kantian thrusts" served equally well as the point of departure for the materialists, proceeding towards the left, and the Machians who derived a rightist inspiration from it.[1]

17, Helmholtz explains that Purkinje's "observations" made in a given experiment were observations of the idiosyncrasies of his own eyes rather than of the eyes he examined. Also cf. PO III 16.

1 "'[V]erschämter Materialismus' mit kantianischen Ausfällen" [W.I. Lenin, *Materialismus und Empiriokritizismus*, in *Werke*, Bd. 14 (Berlin, 1962), pp. 233 ff.].

Yet Helmholtz's views are not just a tedious repetition of familiar themes in the history of ideas. As we will see, one can also witness the dawn of excitingly new insights in his epistemology. Particularly noteworthy among these is his growing awareness of the functional significance of *time* as reflected by his emphasis on the categories of learning and of ontogenetic adaptation as being essentially involved in the development of mature vision and of perception in general. Consciousness is no longer regarded as a passive absorber of phenomenal impressions, as with Locke, nor as a fixed atemporal substance with fixed attributes, as with Kant. Rather, it is hypothesized as *acquiring* the faculties needed for the apprehension of truth through trial and error and by actively engaging in experimental interactions with an as yet unknown reality. The kind of truth to be discovered in this way is indeed no more than a practical one, but the hope of achieving more than this is meaningless:

> In my opinion...there can be no possible sense in speaking of any other truth of our ideas except of a *practical* truth.[1]

Thus the perennial search for the theoretical foundations of knowledge, which has dominated, and still dominates, most epistemological inquiries is gradually making place here, though not yet fully-fledged and self-consciously, for an approach in which knowledge is interpreted as an evolutionary entity and human thought as a peculiar mode of *coping* with reality rather than of merely contemplating it.

---

1 PO III (1962), p. 19.

IX

HELMHOLTZ'S THEORY OF THE PERCEPTION OF SPACE

1. Sensation and Perception.

At the bottom of Helmholtz's epistemological considerations lies an analytic distinction between sensation and perception. It characterizes his epistemological point of view as naturalistic and psycho-physiological. Mere psychological or conceptual analysis leaves one end of the perceptual process as it were dangling in mid-air. One might as well study the life-processes of a tree while leaving its roots out of consideration.

> (There) is little hope that whoever does not start at the beginning, will ever reach the goal of knowledge.[1]

Thus, on Helmholtz's view, a comprehensive psycho-physiology of visual perception largely coincides with a comprehensive theory of (perceptual) knowledge. It should comprise the dioptrics of the eye (how is external light transmitted through the optical apparatus), the physiology proper of optical impressions (how does light, once transmitted, affect the nervous apparatus to produce luminous sensations), and the psychology of visual perception (how do sensations produce perceptions of external objects).

> The physiology of the senses constitutes a border area where the two great divisions of human knowledge—commonly distinguished under the names of the natural and the social sciences—mutually overlap and gear into each other; where problems arise in which both take an equally lively interest, problems, indeed, which can only be solved through cooperative effort on both sides.[2]

---

1 "(Es) ist wenig Aussicht, dass zum Ziele der Erkenntnis kommen wird, wer nicht mit dem Anfang anfängt." (VR I 268).

2 "Die Physiologie der Sinne bildet ein Grenzgebiet, auf dem die beiden grossen Abtheilungen menschlichen Wissens, welche man unter dem Namen der Natur- und Geisteswissenschaften zu scheiden pflegt, wechselseitig in einander greifen, wo sich

Sensations as such are merely mental correlates of internal states of activated parts of the afferent nervous system. Perceptions, by contrast, are states of consciousness in which the sensations are interpreted as signs of an independent reality. Ordinarily we are not even aware of our sensations but only of the perceptions they yield, of the existence, size and location of *objects*. This translation, according to Helmholtz, necessarily involves 'lower psychic activities'.[1] Seeing does not consist in a peculiar excitation of the optic nerve but in the *interpretation* of such an excitation.

Helmholtz's terminological usage, however, is not always consistent. Sensations (*Empfindungen*) are sometimes mere neurophysiological states, at other times they are primitive psychological data. This equivocation, however, is understandable in view of Müller's principle. Thus sensations can be said to be immediate perceptions when they are regarded as the direct psychological results of the excitations of any given nerve. The *opticus* just produces luminous sensations (and nothing else), which as such contain no elements derived from memory of previously experienced sensations. Here the perceived sense-impression and the act of perception are indistinguishable for our consciousness. A higher degree of consciousness is involved, however, when the sense-impressions are not just perceived but *apperceived*, as Helmholtz expresses it in Leibnizian terminology. In the latter case the current sense-impression is instantaneously joined by the idea (*Vorstellung*) of the object in question as retained in memory, that is, by the totality of previously received relevant sensations. Since the apperceptions (*Anschauungen*) of sensory stimuli generally arise in our consciousness no less immediately as do the direct perceptual correlates of sensation, it is often exceedingly difficult to distinguish what in our perceptions of reality is due directly to sensation and what, on the other hand, is the result of experience, training and of unconscious manipulation of the sensory input.

In fact, we will see that Helmholtz develops the outlines of a truly modern information-theoretical account of perception involving as a primitive con-

---

Probleme aufdrängen, welche beide gleich lebhaft interessiren, und welche auch nur durch die gemeinsame Arbeit beider zu lösen sind." (VR I 267).

1 Thus he writes:

> "...to many physiologists and psychologists the connection between the sensation and the conception of the object usually appears to be so rigid and obligatory that they are not disposed to admit that, to a considerable extent at least, it depends on acquired experience, that is, on psychic activity. On the contrary, they have endeavored to find some mechanical mode or origin for this connection through the agency of imaginary organic structures." (PO III 5).

cept his doctrine of 'unconscious inferences'. This theory had great potential as an explanatory principle not only in physiology but also in experimental psychology. Thus, since the mind is utilitarian, impulses that do not deliver useful information are likely to be suppressed altogether unless called up by a special act of the will. Thus we *normally* do not hear partial tones, do not see blind retinal spots, nor two visual fields (since the only useful function of binocular vision is to facilitate the perception of depth), etc.[1] On the other hand, Helmholtz's information theory of perception also enables us to make sense of the formerly incoherent notion of *unsensed sensations*, sensations, that is, which constitute familiar information subconsciously processed. The fact that habitual information processing, if attended to, can reveal sensations previously unnoticed is experimental evidence for the legitimacy of the notion of unsensed sensations. Helmholtz offers many persuasive illustrations which

> indicate that we are exceedingly well trained in finding out by our sensations the objective nature of the objects around us, but that we are completely unskilled in observing the sensations *per se*; and that the practice of associating them with things outside of us actually prevents us from being distinctly conscious of the pure sensation.
>
> This is true also not merely with respect to qualitative differences of sensation, but it is likewise true with respect to the perception of space-relations.[2]

It has been this difficulty, mainly, that has triggered off one of the sharpest controversies among 19th century physiologists. The defenders of the so-called *empirical theory (empiristische Theorie)*, of which Helmholtz was the chief exponent, advocated on both methodological and theoretical grounds a position which allowed as much scope as possible to the role of experience and adaptive development in the complicated psycho-physical relationship between the sense-impressions and our apperceptions of the external world, especially with regard to the latter's spatial attributes. The opposite view, the so-called *intuition theory (nativistische Theorie)* as developed in various directions by Müller, Panum and—in its most extreme form—by Hering, held that the immediacy and urgency, the striking uniformity and universality of all apperceptions alike including the apperceptions of space in general and of localization in particular indicated beyond reasonable doubt that innate factors must be (largely) responsible for the formation of these apperceptions.

---

1 Cf. R.S. Turner, 'Hermann von Helmholtz and the Empiricist Vision', *J. Hist. Behav. Sci.*, 13 (1977), p. 50.

2 PO III 9.

In the heat of the debate the two conceptions appeared to be more antagonistic than they really are. Both have to admit the cooperation of experiential as well as innate factors in the perceptual process. In fact, Helmholtz made sophisticated rhetorical usage of this systematic overlap by exploiting it to the advantage of his own theory in the following argument:

> Then think of the whole system of localization, which, according to Hering, is given originally by direct space-sensation. After the theory has been amended and improved in all sorts of minor ways so as to adapt it better to actual conditions, the most we could ever do would be to make it give a correct localization of objects for a single position of the lines of fixation. In all the innumerable other cases it would be more or less wrong and would have to be amended by experience. Thus Hering's hypothetical assumptions—possibly—do make it easier to explain the visual perceptions in one single instance, by making it all the harder to explain them in every other case. And, at any rate, the conclusion must be that if the factors derived from experience are able to give the correct information as to the relations of space even in spite of opposing direct space-sensations, they must be still better and more easily able to give the correct information about them when there are no such obstacles to be overcome.[1]

There is an unmistakable streak of irony in the last remark. It is even more amusing when Helmholtz, suddenly discomforted, it seems, by the caustic character of the debate, appeals to scientific objectivity in a footnote, thus, by implication, adding rebuke to refutation:

> I have been obliged to make this criticism of Mr. E.Hering's views for the sake of the facts of the case, but I trust it will not be regarded as an expression of personal irritation on account of the attacks which he has made on my latest articles.[2]

Of course, the crucial question at stake in the controversy is not so much concerned with the exact moment at which we can assert the existence of a perceptual system fully equipped for the performance of the complicated tasks on which the life of the organism depends. To select the moment of birth would be rather arbitrary anyhow. On the other hand, if the intuitionists allow the hypothesis that the perceptual faculties may partially develop over time, then what remains to be discussed is the origin and the nature of this development. If immediacy is taken as the decisive criterion of all bodily action upon the

1 PO III 557

2 Ibid., n. 1.

mind while the higher psychic functions, especially those involved in making judgments, are regarded as being of an altogether different character, obeying as they do the 'laws' of mental life, then the determination of the nature of any given psychic manifestation at once decides the mode of its origin. Accordingly all spatial perceptions in particular, on account of their apparent immediacy, must be classified as sensations and the laws of their development must be genetically fixed laws of physiological growth rather than laws of psychological development and of trial-and-error learning. Thus, as von Kries points out, the central question underlying the controversy is

> whether the spatial determinations of our sensory impressions can be differentiated from their other (qualitative and intensive) determinations as something that is fundamentally different—different partly as to their psychological nature, their inner relationships etc., and partly as to their genesis, the nature of the mechanism by which they are produced, and especially their dependence on experience.[1]

2. The General Idea of Space and Perceptual Localization.

Kant had argued that space is not an objective property of the universe which we can perceive through the senses. Rather, the idea of space is an a priori form of perception which constitutes the objects of perception given by experience. Moreover, the antinomy of space inherent in the notion as applied to the universe exposes the ultimate limitation of our mental powers to describe the world. For the universe must be finite on the one hand, unbounded on the other, yet representable in the third place. However—in the language of laboratory models—an exact scale model preserving the metric features of a universe which is both finite and homogeneous is logically impossible. Riemann later showed that a finite unbounded universe *was* conceptually possible. He provided in fact the geometric language for Einstein's relativistic physics, which of course went far beyond Riemann's mathematical results. What is relevant here is the further fact that a true model of the galactic system does not necessarily have to be exactly scaled. We can filter out the extrinsic geometry of space while capturing in our model its intrinsic features. Gauss had already discovered that the curvature of a surface can be

---

1 PO III 650.

completely determined solely from the distance distortions present in a chart. Hence an atlas of maps is a true model of the earth's surface, its spatial distortions representing the intrinsic property of the positive curvature of that surface. Quite similarly, the completely analogous distortions in Einstein's model of the galactic system are attributable to the curvature of space.[1]

Kant listed the following a priori characteristics of our notion of space. It is impossible to conceive of the absence of space, although it is perfectly easy to imagine that no objects are encountered in it.[2] The idea of space, therefore, is *invariable*, unlike the various qualities of sensation which an object may or may not have without ceasing to be an object of perception. It is also a *homogeneous* idea. It is impossible to conceive of an individual place all by itself. To be sure, in some sense the perception of most qualities of sensation is indeed contextual as well, yet the connection here is not necessarily 'environmental' as is the connection between the perception of definite places on one hand and the general idea of space on the other.

Kant's transcendental theory, however, though it explained the attribution of spatiality *in general* to whatever is the object of human perception, left unresolved the question as to how *specific* and *definite* place-attributions are possible. It described the synthetic functions of our consciousness which are indispensable for the very possibility of any spatial interpretation of what is sensed in the first place, but it shed no light whatsoever on the processes governing *localization*.

> Even Kant, who for us coming after him brought to a final conclusion the earlier attempts at a theory of knowledge, still compressed into one single act all links between pure sensation and the formation of a representation of a given object in time and space. This act he called *intuition*. With him and his successors this act plays a role as though it were in its entirety the immediate effect of nothing more than a purely natural mechanism, not further analyzable by psychological or philosophical investigations except for its final result which is precisely a representation and thus subject to certain formal conditions of representation as such which Kant identified.[3]

1 Cf. J.J. Callahan, 'The Curvature of Space in a Finite Universe', *Sci. Am.*, (Aug. 1976), 23, 2, pp. 90-100.

2 Imagine how stunned poor old Descartes would have been upon learning this proposition.

3 "...[S]elbst noch Kant, der für uns Nachkommende das Facit aus den früheren Bemühungen der Erkenntnistheorie gezogen hatte, fasste noch alle Zwischenglieder zwischen der reinen Sinnesempfindung und der Bildung der Vorstellung des zur Zeit

However, even if *spatiality* is a necessary and a priori form of outer perception, as Kant had urged, it does not follow that our perceptions of *locality* cannot have been acquired. In fact, this latter view was defended first by Lotze, on the basis of psychological considerations, and then by Helmholtz, on the basis of extensive empirical research.

In order for definite place-ascriptions to the qualities of sensation to be possible at all, Lotze argued, the impressions on the retina (or in the terminal fibers of any other, e.g. the cutaneous, nerves) must be accompanied by "local signs" which enable us to discriminate between sense impressions on different parts of the retina. The assumption that the bare existence of a retinal image provides us with direct information about the spatial arrangement of its parts and suffices to explain the immediate localization of the corresponding external objects is neither necessary nor plausible, according to Lotze. Contrary to the largely *empirical* arguments Helmholtz was later to adduce in support of this assertion, Lotze's psychological argument is strictly *formal*. As it turns on the essentially non-spatial, non-extended nature of the *sensorium commune* or the soul, it is as convincing as the Cartesian view it presupposes:

> The mental presentation *is* not what it presents; and the idea of a point on the left does not lie on the left of the idea of a point on the right.[1]

The clues utilized in the formation of conscious perceptions for the purposes of localization are accessory impressions which are no more than empty, non-spatial signs (*Localzeichen*) whose intensity varies as a continuous function of the stimulated place of the receptor-organ, while the other qualitative and intensive determinations of sensation do not systematically vary with such places. Thus, with regard to localization as afforded by visual perception,

> all spatial differences and relations among the impressions on the retina must be compensated for by corresponding non-spatial and merely intensive relations among the impressions which exist together without space-form in the soul;

---

wahrgenommenen, räumlich ausgedehnten Gegenstandes in einen Act zusammen, den er die *Anschauung* nannte. Diese spielt bei ihm und seinen Nachfolgern eine Rolle, als wäre sie durchaus nur Wirkung eines natürlichen Mechanismus, der nicht weiter Gegenstand philosophischer und psychologischer Untersuchungen werden könnte, abgesehen von seinem Endergebniss, welches eben eine Vorstellung ist, und also auch unter gewissen formalen Bedingungen alles Vorstellens stehen kann, die Kant aufsuchte." [H. von Helmholtz, 'Goethe's Vorahnungen kommender naturwissenschaftlicher Ideen', (1892), in VR II 338].

1 Hermann Lotze, *Grundzüge der Psychologie*, (Leipzig, 1882), p. 28; quoted in W.N. Dember, *Visual Perception: The Nineteenth Century*, (New York, 1964), p. 133.

> and...from them in reverse order there must arise, not a new actual arrangement of these impressions in extension, but only the mental presentation of such an arrangement in us.[1]

3. The Intuitionist Theories of Müller and Hering.

An entirely opposite view, though in its point of departure no less inspired by Kant than Lotze's, was held by Müller. Under the impression that his physiological principle of specific sense-energies provided some tangible support for Kant's theory of the subjectivity of the phenomenal world, Müller likewise sought to extend Kant's transcendental view of the subjective apperception of space into the physiology of perceptual localization. He conjectured that the retina must be sensitive to its own extension in space, thus affording the perception of subjective spatial arrangements of retinal sensations. The problem of the psycho-physical relationship involved in the special localization of sensory information is thus solved by the assumption of an innate contrivance which makes possible the mind's immediate awareness of the retinal dimensions. For Lotze, contrariwise, the objective spatial configuration of the processes in the organs of sense was *not* a sufficient basis for the development of any subjective spatial order:

> an impression is not *seen* at a definite point on account of its *being situated* at such point; but it may perhaps by means of this definite situation *act* on the soul otherwise than if it were elsewhere situated.[2]

Müller's theory paved the way for Hering's intuition theory of perception (*nativistische Theorie*) which allows little scope for the cooperative role of experience in the perceptual process. Rather, it attempts to derive, through innate mechanisms, the entire content of our perceptions (including their spatial attributes) as immediate results of the sensations produced by the organs of sense. The train of thought initiated by Kant and extended by Müller was transmogrified beyond recognition in Hering's intuitionist physiology which would go, rather, to support a purely sensationalist epistemology. Hering abandoned the notion of the invariable and homogeneous nature of the general idea of space as such. Spatial perception

---

1 Ibid., pp. 30-1; quoted in W.N. Dember (1964), p. 135.

2 Ibid., p. 30; quoted in W.N. Dember (1964), p. 134.

does not require any mediating activities of the mind. Any random aggregate of impressions on the retina yields as an unconditioned reflex the conscious perception of a minutely detailed spatial order among the luminous phenomena of, possibly, an entirely unfamiliar reality through the immediate sensation of definite place and depth values. The fundamental Kantian distinction between the phenomenal qualities constituting the content of immediate experience on the one hand (whatever their 'real' cause), and on the other their apperception within a spatio-temporal matrix which involves an extremely complex cognitive process radically different from mere sensation, is entirely obliterated. Space and place are directly sensed as such, just as any other quality of sensation.

4. Helmholtz's Empirical Theory of Perception.

The arguments Helmholtz advances in support of his empirical theory can be divided into three main categories: theoretical, methodological and philosophical, if not (*pace* Helmholtz) metaphysical, arguments. I will review them in this order. A discussion of Helmholtz's views on the formation of the *general* concept of space will then finally bring us back to his philosophical relationship to Kant.

Helmholtz states that

> the sensations of the senses are tokens for our consciousness, it being left to our intelligence to learn how to comprehend their meaning.[1]

When this assertion is taken to pertain to all sensory inputs including Lotze's so-called local signs (*Localzeichen*), it expresses the fundamental thesis of the empirical theory.

Generally speaking, the spatial distribution of stimulated nerve fibers does not *inevitably* produce the idea of locally disparate causes of these excitations. A smoke-filled, noisy room excites different (e.g. olfactory and auditory) nerve systems without creating the impression of locally separated agencies. Rather, we get the correct impression that these agencies co-exist simultaneously in the air all around us. Similarly, we can see a single object with two separate eyes. In fact, two instances of the same kind of sensation

---

1 PO III 533.

may, under slightly different circumstances providing us with different accessory information, produce spatial perceptions that are radically at variance with each other. Thus, when we touch a table with both index fingers and feel a grain of sand under each of them, the sensation we have is, apart from accessory information concerning the position of our fingers, essentially the same as when we press the index fingers against each other with a grain of sand between them. Yet, in the former case we form the perception of two separate grains while in the latter we perceive only one. It is well known that when we have a false or vague idea of the position of our fingers as when two of them are crossed, we believe to feel two marbles, say, even though we are touching only one between the two fingers.[1]

Hence the question arises what factor *in addition to* the spatial separation of the sensitive nerve fibers must be assumed in order to account for the spatial discrimination in our perceptions. The search for a solution of this question seems to lead inevitably to a hypothesis of local signs. But while the empirical theory considers these signs as arbitrary symbols, devoid of intrinsic spatial meaning and as such indistinguishable from the qualitative and intensive signs of sensory information,[2] the intuition theory presupposed that the local signs are nothing but immediate perceptions of the spatial distinctions as such, with regard to their nature as well as their magnitude.

The most serious difficulties confronting the intuition theory are related to the phenomena of depth perception. Spatial discrimination with regard to (near-)planes in the field of vision may perhaps be explained satisfactorily by both theories. Even here, however, the problems facing the intuitionists are much greater than those facing the empiricists. For the intuitionists need to account in some special fashion for certain incongruities between the retinal and the perceptual images such as the well known inversion of the retinal image as distinct from the perceptual image of reality we obtain, as well as for the fact that the two retinal images combine to produce single vision. In order to solve this second problem they were led to assume that the points on the two retinas which correspond with each other—the correspondence being mathematically defined—are in fact physiologically *identical* points producing single sensations. One might want to seek anatomical evidence for this assumption in the fact that the two optic nerves cross each other in the optical chiasma before they enter the cortical hemispheres, the right hemisphere receiving the fibers from the right retinal halves, the left one those

---

1 H. von Helmholtz, 'Die neuere Fortschritte in der Theorie des Sehens', (1868), VR I 330-1.

2 Ibid., p. 354.

from the left retinal halves. However, this neurophysiological datum by no means constitutes any evidence for the *fusion* of corresponding fibers.[1]

The empirical theory, on the other hand, gives of course short shrift to the puzzles of plane vision. Which points on the two retinas correspond with each other is a matter of experience and the two impressions are accordingly combined in a single perception. Experience, similarly, solves the problem of retinal inversion. The harmonization of the senses of touch and of sight is gradually developed through meticulous learning and through trial-and-error adjustment. Stratton's celebrated experiment, inspired by the empirical research program, of prolonged stimulation by retinal images that are not inverted seems to indicate, moreover, that continuous re-harmonization remains possible though requiring considerable effort.[2] This experiment thus provides corroborating evidence for the Helmholtzian research program.

But let me now turn to the phenomena of depth perception as conditioned by binocular vision. How can impressions of corporeal three-dimensional objects be generated in perception on the basis of two plane retinal images which are perspectively different? Wheatstone's invention of the stereoscope in 1838 greatly accelerated research in tridimensional localization. Before him, the utilization of various cues in the perception of distance had been recognized or hypothesized. The important influence of certain so-called 'empirical' factors such as linear and aerial perspective, interposition and the like was already well-known to painters of the late medieval and early Renaissance period, with such artist-(engineers) as Paolo Uccello, Piero della Francesca, Brunelleschi, Alberti and Leonardo da Vinci being the main pioneers in the theory of perspective representation. Other, physiological, cues had also been proposed such as the proprioceptive information concerning the muscular efforts involved in accommodation (the changing of the lens shape in accordance with object distance) and in convergence (ocular rotations of the two eyes in concert about the horizontal axis for the maintenance of direct fixation on objects regardless of their distance).[3] Wheatstone's experiments, however, made a very strong case for the view that the single most important cue utilized in the perception of distance and of relative depth is *retinal disparity* produced by binocular parallax. This meant that the elucidation of the notion of corresponding retinal points became the focal point of the controversy between intuitionists and empiricists.

---

1 Ibid., p. 334.

2 G.M. Stratton, 'Vision without Inversion of the Retinal Image', *Psych. Rev.*, (1897), 4, 341-481.

3 Cf. PO III 54-8; 281-93; 604.

Fig. 19 — Some monocular cues for the visual perception of depth.

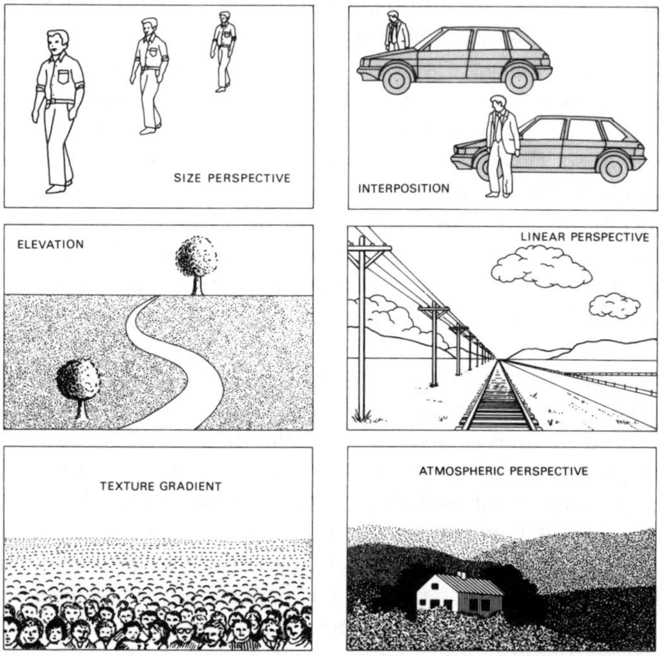

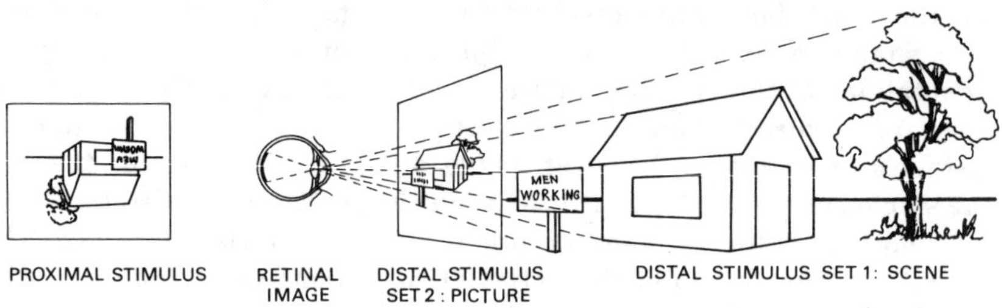

Fig. 20 — Illustration of interposition. *The diagram demonstrates that the 'cues' of interposition can be projected as retinal images by more than one, indeed by infinitely many, sets of distal stimuli (in this case the real scene and a two-dimensional picture thereof).*

Do we *learn* which points on the retinas correspond with each other or is the correspondence *automatic*, perhaps even anatomically fixed? Since this question cannot be decided by direct observation, its solution must depend on theoretical considerations.

If localization is determined strictly by innate capabilities, then we must also be originally endowed with a faculty of divining what points on the two retinas are corresponding or, in the intuitionists' terminology, 'identical' points. Corresponding points are those respective points on the retinas or their outward projections, the so-called visual globes or fields, that are equidistant and in the same lateral or vertical direction from the point of fixation, and whose images, therefore, coincide in the common field of view. It is a necessary consequence of the intuition theory that such points are *identical*, their identity usually being hypothesized as fixed by anatomical structures in such a way that sensations originating from any pair of corresponding points will fuse into one single impression. But, as Helmholtz will argue, it is precisely this basic presupposition of the intuitionists which will prove to be untenable in the face of the various facts of depth perception, thus rendering their research program implausible or even ill-conceived.

First of all, we do not ordinarily discern very clearly those double images that can be smoothly combined into the perception of extended bodies, whereas the perception of stereoscopic relief occurs with astonishing accuracy. In the latter case the slightest disparity is immediately converted into the stereoscopic perception of depth. This, incidentally, explains why it is so easy to identify counterfeit bank notes simply by comparing them under a stereoscope with genuine notes. Since, however, stereoscopic perception utilizes the very same differences between the retinal images which underlie the appearance of double images, the discrepancies between the two kinds of perceptual responses cannot be adequately explained in terms of retinal disparity alone.

Moreover, by varying certain extraneous conditions we can manipulate (impede or enhance) the readiness with which we discern the same double images. However, such conditions should have no influence if localization were determined by some innate mechanism, whether an anatomical connection of corresponding nervous conductors or an automatic instinctive fusion of their respective sensory information.[1]

1 Cf. VR I 343-5.

Further difficulties arose in connection with retinal disparity as a cue for depth perception. At first, Brücke almost succeeded in reconciling the phenomena of binocular perception with the theory of innate retinal identity by drawing attention to the important factor of ocular movement by means of which the eyes continuously scan the various contour lines in the field of vision.[1] He suggested that depth perception might result from the proprioceptive gauging of the muscular efforts involved in ocular movement which vary with the distance of the point of fixation. According to Helmholtz, however, this hypothesis was effectively refuted when Dove demonstrated that the peculiar illusions generated by stereoscopic pictures are sustained even under instantaneous illumination by an electrical spark, that is, during a time span too short for the occurrence of any ocular movement.[2]

Yet, Helmholtz's methodological evaluation of this experiment may seem to be a little rash and naïve. For he takes Dove's experiment to be a crucial one, with all the classical failures inherent in this notion. Indeed, as a matter of historical fact conflicting hypotheses were built upon Dove's experiment. Thus Panum even argued, quite to the contrary, that Dove's experiment demonstrated that space is innate and not generated by eye-movement.[3] However, in order to maintain this Panum was forced to considerably weaken the original theory of identical retinal *points* and to introduce instead a new *ad hoc* notion of corresponding sensory *circles*. This notion was based on Weber's conception of sensory circles on the skin. Weber held that double stimulation within a given cutaneous sensory circle would be perceived as single. Panum exploited this theory to argue that similar contours falling within corresponding sensory circles on the two *retinas* would likewise be perceived as single. From this point he went on to discuss the "sensations of binocular parallax" that could give rise to the perception of depth and to a non-kinæsthetic visual space. Hering's nativistic theory of visual space was influenced greatly by Panum. Ultimately this line of theorizing would lead to the nativism inherent in Gestalt psychology and to the notion (expressed in Köhler's theory of cortical isomorphism) that the properties of the perceptual field depend upon an organization of phenomenal data which is physiologically fixed.[4]

---

1 E. Brücke, 'Über die stereoskopischen Erscheinungen', *J. Müllers Archiv für Anatomie und Physiol.*, (1841), p. 459.

2 H.W. Dove, *Monatsber. d. Berl. Akad.*, (July 29, 1841).

3 P.L. Panum, *Physiologische Untersuchungen über das Sehen mit zwei Augen*, (Kiel, 1858); id., 'Über die einheitliche Verschmelzung verschiedenartiger Netzhauteindrücke beim Sehen mit zwei Augen', *Reicherts Arch. für Anat. u. Physiol.*, (1891), pp. 63-111, 178-227.

4 Cf. E.G. Boring (1942), p. 288.

Dove likewise discovered the phenomenon of stereoscopic gloss produced by different coloring of the two stereoscopic pictures.[1] The fact that black and white in stereoscopic combination yield the perception of a lustrous surface instead of one that is dull gray entails that the impressions of the two retinal images do *not* fuse into a single sensation. Nor is it possible to invoke the so-called phenomenon of retinal rivalry, discovered by Wheatstone, in order to explain the stereoscopic perception of gloss. For, again, not only does the appearance of luster persist even under instantaneous illumination, but it is also demonstrably enhanced precisely by the *prevention* of rivalry. Moreover, the intuitionists' appeal to rivalry in this case is theoretically inconsistent inasmuch as the phenomena of rivalry between the visual globes are related to the *attention*, that is, they originate in *psychic* activities rather than being produced by physiological processes.[2]

More positively, it can also be shown that the two impressions we receive from corresponding retinal places are systematically discriminated. If they were fused anatomically or instinctively, then it would have to be immaterial which eye is shown which picture. However, when we exchange the stereoscopic pictures accordingly, the resulting difference in perception is, on the contrary, quite drastic! For when the right eye is shown the left picture and the left eye the right picture, we obtain a *pseudoscopic* impression in which the original relief is reversed. Yet, even at instantaneous illumination of a stereoscopic line drawing we always perceive its correct relief, whereas on the intuition theory we should obtain the reverse relief just as readily and frequently as the correct one.[3]

Finally there are the cases of strabismus and of various other anomalies of ocular adjustment. They seem to show that the primary relations of correspondence can be modified and reorganized into entirely new systems of secondary correspondence. There could hardly be any better argument in favor of the theory which attributes the formation of *any* relations of correspondence to training and experience. Correspondence is a *functional* relation, not a rigid built-in structure.[4]

---

1 H.W. Dove, 'Über die Ursache des Glanzes und der Irradiation, abgeleitet aus chromatischen Versuchen mit dem Stereoskop', *Poggendorffs Ann.*, 83, (1850), p. 169.

2 VR I 346-8.

3 Ibid.

4 Cf. J. von Kries, PO III 579/592.

5. Methodological Arguments in Defense of the Empirical Theory of Perception.

Helmholtz admits quite frankly that the present state of neurophysiology does not suffice for a decisive refutation of the intuition theory.[1] In fact, he is at times surprisingly generous in his recognition of this fact. Thus he states:

> It ought to be said in the beginning that our knowledge of the relevant phenomena is still too limited to justify us in accepting *any one theory* to the exclusion of all the others.[2]

Elsewhere, after having summed up a number of general considerations which count against the plausibility of the intuition theory, his conclusion is again very modest and quite acceptable:

> This is by way of justifying my point of view. A choice had to be made simply for the sake of getting at least some sort of superficial order amid the chaos of phenomena; *and so* I believed I had to adopt the view I have chosen.[3]

However, immediately after having declared that at the present state of physiological inquiry we are not justified in accepting any one theory of perception to the exclusion of all others, Helmholtz slowly begins clearing the ground for the sole acceptance of his own theory. He appeals to scientific sobriety and deplores the

> tendency...to yield too much to a predilection for certain metaphysical modes of thought, instead of being guided simply by the facts themselves; especially as fundamental questions continue to arise in the realm of psychology which have long since been completely barred from the domain of the phenomena of inorganic nature.[4]

But the intuitionists, of course, had offered more cogent arguments than the mere metaphysical accusation that Helmholtz's introduction of psychic activities as mediating in the perceptual process smacked of idealism. Their whole point had been that the psychological assumption was first of all superfluous in view of their own physiological explanations[5] and, secondly, implausible in view of the ubiquitous phenomenon of highly accurate spatial perceptions by newly-born animals, which, except for the theoretical

---

1 PO III 17/531.

2 PO III 531; emphasis added.

3 PO III 18; emphasis added.

4 PO III 531.

5 This argument can be turned around, of course, with equal force; cf. PO III 17.

possibility of prenatal experience, must be assumed to be entirely instinctive. On the other hand, Helmholtz himself at one point flatly asserts that in his view it is entirely inconceivable that detailed spatial perceptions can arise immediately, without any mediation of the mind and unaided by any previous experiences. While this may be a sound intuition of course, it is certainly not one that is valid a priori. If presented as an argument, therefore, its force, if any, can be no more than metaphysical. And this Helmholtz himself is subsequently forced to concede. Thus he writes:

> The first objection which I should have to offer, and which to my way of thinking seems absolutely insuperable, is that I cannot conceive how a single nervous stimulation can produce a completed idea of space without antecedent experience. However, I realize that this objection is probably of too metaphysical a nature to be considered on scientific grounds, and so I simply register it here for the benefit of those readers who share my view.[1]

The next step in Helmholtz's argument abounds with very clever, possibly unintended, innuendos. By ascribing to "natural philosophers" the very same theoretical inclinations he has previously taken his scientific opponents to task for, he implicitly puts the latter in the same boat as the speculative philosophers who had dominated the universities of Germany ever since Hegel and whose intellectual enterprise had only recently become anathema among the more experimentally minded scientists.

> Many natural philosophers have been far too ready to presuppose all kinds of anatomical structures in the theory of visual perception and also to postulate new qualities of the nervous substance that are contrary to what we actually know about the physical and chemical properties of bodies in general and about the nerves in particular.[2]

There is also something of methodological interest in this quotation. For Helmholtz implicitly espouses here a methodological rule which stems from a Newtonian doctrine whose influence is still widely felt. It stipulates that no novel hypothesis is admissible which is incompatible with already accepted and therefore more firmly entrenched hypotheses.[3] The doctrine serves to

---

1 PO III 551.

2 PO III 531.

3 Cf. Newton's notorious Rule IV of the *Regulae Philosophandi* at the beginning of Book III of the *Principia*, added only since the third edition of that book.

institutionalize the worst kind of theoretical conservatism.[1] This one can acknowledge even without committing oneself to the extreme opposite view of the epistemological anarchist who condemns all methodological legislature and would rather encourage than prohibit the daily overthrow of the whole of physics. Clearly we need a minimal degree of theoretical stability. And thus we are justified, if only pragmatically, in accepting some fundamental theories as the premises for further research. These theories, however, express certain general 'facts' about the world only because, for the time being, we have legislated against the very possibility of their falsification. As such they constitute the negative heuristic cores of our current research programs.[2]

On the other hand, if we are to identify certain experimental results as potentially falsifying currently accepted hypotheses, then we clearly need alternative theories incompatible with the hypothetical 'facts' of the day. Failing such theoretical competitors no comprehensive theory can ever be overthrown. As long as we are incapable of interpreting, i.e., of making sense of, certain recalcitrant facts we are compelled to throw them into the wastebasket of annoying anomalies rather than glorify them as so-called crucial experiments.[3]

But let us explore Helmholtz's argument a little further. He quite skillfully defends his methodological conservatism, and consequently—or so it seems at least—the superior viability of his research program as resting on more solid foundations. The reasons he offers (assuming the truth of what they assert) can hardly fail to persuade us.

> It is safer, in my opinion, to connect the phenomena of vision with other processes that are *certainly present* and *actually* effective, although they may require further explanation themselves, instead of trying to base these phenomena on *perfectly unknown hypotheses* as to the mechanism of the nervous system and the properties of the nervous system and the properties of the nervous tissue, which have been *invented for the purpose* and have *no analogy of any sort*. The only justification I can see for proceeding in this way would be after

---

1 Cf. P.K. Feyerabend, 'How to be a good Empiricist—a Plea for Tolerance in Matters Epistemological', repr. in P.H. Nidditch (ed.), *The Philosophy of Science*, (Oxford U.P., 1968).

2 Cf. I. Lakatos (1970).

3 This point was first and most forcefully made by Paul Feyerabend; cf. op. cit., in Nidditch (1968).

> all attempts had failed to explain the phenomena by ***known facts***.[1]

But what then, we are inclined to ask, are these so-called "known facts"? And what are these processes we are invited to consider as "certainly present" and as "actually effective"? As it turns out, these are the psychic activities which according to Helmholtz's empirical theory of perception are indispensable for the formation of correct spatial interpretations of the optical (tactile, etc.) sensations we have.[2] But are we than to regard these psychological processes as "known facts"? That certainly may strike us as a rather baffling proposition. For in the first place, it seems to me, we need psychology, that is, a systematic investigation of these mental processes in order to elevate our muddled awareness of them to a more or less respectable state of theoretical knowledge. Most of us are not ordinarily in the habit of reflecting upon our mind's operations, let alone of musing upon its "laws".

Secondly, even granting Helmholtz's insistence that "to a certain extent we are familiar with [such] laws from daily experience"[3], we must bear in mind that the knowledge Helmholtz graciously attributes to us cannot simply concern the generally acknowledged *existence* of psychic phenomena and their regularities but rather their peculiar *interactions* with the data of sensory experience. However, of this "absolutely unquestionable interplay of psychic phenomena"[4] which the natural philosophers "either ignore entirely...or dismiss...as of comparatively slight importance"[5] we *knew* absolutely nothing until Helmholtz's theoretical exertions may now finally have opened our eyes. That is, what are supposed to be known facts are actually highly theoretical assertions which constitute the very bone of contention between the empiricists and the intuitionists. The waves of irony go very high indeed, when Helmholtz goes on to say of the *intuitionist* hypotheses, which, as Helmholtz has managed to imply, share the same theoretical foundation with those of the "natural philosophers", that "while they are intended to have the appearance at least of being scientific explanations"[6], they clearly disqualify for that honorific simply because they ignore the "absolutely unquestionable"[7] psychological facts.

---

1 PO III 532; emphasis added.
2 Cf. PO III 531-2.
3 PO III 532.
4 Ibid.
5 Ibid.
6 Ibid.
7 Ibid.

How profoundly speculative Helmholtz's psychological assertions really are may even more clearly dawn upon us when, in the third place, we consider the fact that the psychic activities involved in any given perception are in his view *unconscious* mental processes of which, by definition, we have no direct knowledge whatsoever. His psychological assumptions, therefore, far from embodying *solid facts* upon which he erects his empirical theory of perception, constitute on the contrary its ultimate *theoretical pinnacle*. His intriguing theory of unconscious inferences, which was highly controversial from the very beginning, helped pave the way for entirely novel psychological explorations. Thus the profound irony of the matter is that while posing as the steadfast theoretical conservative and while chiding his scientific opponents for methodological vagrancy, Helmholtz turns out to be the boldest innovator launching the most daring hypotheses while cleverly disguising them as 'facts'.

There is one more apparent irony worth mentioning. Helmholtz has criticized the intuitionists for making unwarranted assumptions about "all kinds of anatomical structures"[1] and about new properties of nervous tissue contrary to what was "known"[2] about it at the time. However, given his longstanding acceptance of the idea adopted from Müller that the brain is the organ of the mind and the center of our volitions, he was consequently compelled to concede that the lower psychic activities involved in the development of conscious perceptions must themselves be founded on a neurophysiological substratum, nay, might consist in no more than certain material brain processes beyond our immediate control. Thus Helmholtz's own research program, which we were urged to favor over that of the intuitionists with their narrow, and hence 'unscientific', attempts at purely neurophysiological explanations of perception, ultimately flows back, it seems, via a detour through psychology, into the mainstream of neuro-physiological inquiry. One might as well insist on canalizing a river by *adding* meanders to it.

Of course it is not quite accurate to imply that we are now round full circle. The intuitionists focused on explanations within the narrow confines of physiological *optics*. Helmholtz's apparent willingness to accept a physiological approach towards the study of learning processes would rather designate the *physiology of the central nervous system* as the proper field of inquiry. The essential difference that remains between the two theoretical perspectives may be, then, that on Helmholtz's view the neurophysiological circuits with which our sensory system is originally endowed are subject to further devel-

1 PO III 531.
2 Ibid.

opment in the course of the organism's maturation, not through mere organic growth but rather through a process of *flexible* trial-and-error responses for the purposes of optimal adaptation.

> If anyone objects to including these processes of association and the natural flow of ideas among the psychic activities I will not quarrel over names. Possibly the empirical theory might be united here with that form of the intuition theory which was proposed by Panum, for instance, except for the fact that what he regards as being natural endowment appears to me to have been acquired by experience.[1]

6. The Philosophical Significance of the Intuitionist-Empiricist Controversy.

For Helmholtz what was at stake, ultimately, in the controversy regarding the physiology of perception was a thesis of hypothetical scientific realism which, as we shall see, he was at pains to defend throughout his epistemological expositions. The *real* world, as we shall see, not just the phenomenal one, can be known after all, although this knowledge rests on ultimately unverifiable hypotheses. The beginning and end of all knowledge, however, is *practical*, and this fundamental practice through which we learn how to cope with reality begins not in the laboratory but in the nursery room.

The greater the adaptive potential in the individual's lifetime, the greater the possible range of correspondence between our cognitions and varying environments. The very possibility of intelligent behavioral interaction with the environment is conditioned by the flexibility of the processing devices for sensory inputs. Thus, paradoxically enough, Helmholtz's realism demanded on the physiological side a *lesser* degree of fixed information processing structure in any given sensation. On the psychological side, by contrast, his thesis of pragmatic realism led him to the recognition of proto-intellectual activities and made him clear the way for the field of functional psychology.

If our spatial perceptions were generated in full-fledged form by mere mechanical responses of innate physiological structures, there would be no room left for any interpretative activity. Since sensations cannot be overridden by

---

1 PO III 541.

experience, our localizations as allegedly afforded directly by the senses cannot have been gradually attuned to reality by a process of trial and error. On the other hand, they are the *effects* of external causes and hence essentially different from the latter. Thus one must either deny the existence of any fit between our perceptions and reality, or else one is compelled to take recourse to the metaphysical doctrine of pre-established harmony, which, however, failing any corroborating scientific evidence amounts to no more than a gratuitous article of faith.

This latter assertion is not necessarily true of course. But neither Helmholtz nor his opponents ever sought to invoke a Darwinian model of explanation for the supposed harmony. Obviously, the eye can be studied from a macro-evolutionary perspective as well. It has evolved as a very powerful substitute for the more primitive and cumbersome methods of locomotor exploration. Against such a theoretical backdrop the entire controversy between Hering and Helmholtz would appear to be futile. For if the emphasis on the high relative contribution of experience versus innate factors to the final products of perception should mainly serve to buttress the philosophical hypothesis of scientific realism, then—on the strength of the structural similarities between individual learning on the one hand and molar processes of selective retention of adaptive variations on the other—this objective could equally well be served by stressing the importance of physiological contrivances having evolved, and thus inductively tested, not so much during the individual's life time but rather over long periods of successive generations of the species. Helmholtz states that

> the pervasive contrast between various philosophical systems which either presuppose a pre-established harmony between the laws of thought and of conception and the external world, or contrariwise seek to derive all such agreement from experience.[1]

has its ramifications all the way down into the physiology of spatial perception. From our modern vantage point we can now recognize that the debate is not philosophically relevant in precisely the sense Helmholtz envisaged.

But to belittle its importance as lacking epistemological relevance altogether, as Campbell implies[2], is equally unjustified. Not only because the dividing line between pure philosophy and empirical science rests itself on a dubious

---

1 VR I 333.

2 Donald T. Campbell, 'Methodological Suggestions from a Comparative Psychology of Knowledge Process', *Inquiry*, 2 (1959), p. 160.

philosophical decision.[1] And again, not only because traditional epistemology is progressively giving way to the "psycho-bio-sociology of cognitive behavior"[2] to whose theoretical developments philosophers can contribute substantially, as Campbell himself points out.[3] But also because the theoretical issue concerning innate endowment versus training and conditioning in the perceptual process bears heavily on such outstanding epistemological questions regarding the origin and character of man's synthetic activities, the nature of the psycho-physical relationship, the meaning of the notion of observability (still an "undefined primitive term" in Popper's philosophy of science), and the effective range of influence of subliminal inferential processes. Especially with regard to this latter point Helmholtz rightly emphasizes the contribution his physiological theorizing has made to opening up new avenues of psychological research and of epistemological reflection:

> I believe we must regard the analysis of the concept of intuition into the elementary processes of thought, an analysis still failing in Kant, as the most essential progress of the modern era.... Due to physiological investigations into sensory perception in particular we have been brought to the ultimate elementary processes of thought, processes inexpressible in words and thus doomed to remain incognizable and inaccessible to philosophical inquiry as long as the latter restricted itself to an analysis only of knowledge as expressed in language.[4]

Helmholtz has not entirely ignored, though, the evolutionary point of view. In fact, in a few occasional remarks he comes surprisingly close to a full recognition of its impact on (neuro-)physiology as well. Discussing the laws of ocular movement he drops a parenthetical remark which, but for its lack of elaboration, may seem to suggest a rather novel idea:

---

1 Cf. VR II 244.

2 H. Feigl, 'Existential Hypotheses: Realistic versus Phenomenalistic Interpretations', *Phil. Sci.*, 17 (1950), pp. 35-62.

3 Campbell (1959), pp. 155 f.

4 "Als wesentlichsten Fortschritt der neueren Zeit glaube ich die Auflösung des Begriffs der Anschauung in die elementaren Vorgänge des Denkens betrachten zu müssen, die bei Kant noch fehlt... Es sind hier namentlich die physiologischen Untersuchungen über die Sinneswahrnehmungen gewesen, welche uns an die letzten elementaren Vorgänge des Erkennens hingeführt haben, die noch nicht in Worte fassbar, der Philosophie unbekannt und unzugänglich bleiben mussten, so lange diese nur die in der Sprache ihren Ausdruck findenden Erkenntnisse untersuchte." (H. von Helmholtz, 'Die Tatsachen in der Wahrnehmung', SE 134; EW 143).

> ...so far from its being impossible, I am inclined to think that it is even probable that the growth of the muscles, perhaps too even the efficiency of nervous transmission, is so adapted to the demands made on it during the life of the individual, and perhaps by inheritance during the life of the species, that the requisite movements that are the most suitable become also the easiest to execute.[1]

One cannot help associating this lucky hunch with Anton Dohrn's principle of "*Funktionswechsel*" (1875), according to which changes in organic structure represent accommodations to functional shifts as initiated by the *behavioral* changes of the organism and in conformity with its subsequent experiences. Thus behavior should be the evolutionary pacemaker.[2]

At any rate, Helmholtz's main objection to the assumption of innate contrivances—apart from the fact that their existence can only be postulated, not actually shown—is that these assumptions do not *suffice* to explain the phenomena. For whatever can be overcome by experience or modified by the will cannot in its entirety be due to the blind operation of instinctive response mechanisms. And since—in the case of ocular movements for instance—

> it is possible to show that, at the instigation of voluntary effort, exceptions can occur in case of all these laws of ocular movements, when there are optical ends to be gained by it..., these laws cannot be dependent on anatomical contrivances that act mechanically.... In any event, this anatomical mechanism, if it exists at all, is merely conducive, and not obligatory.[3]

7. The General Idea of Space.

Helmholtz not only claimed that localization is a skill that we have mastered through practice, he also held that the *general* category of space as such is essentially derivable from experience. At first we only observe that through the innervation of certain motor nerves we can bring about changes which we perceive through the senses of touch and of sight. Indeed, we are not aware of

---

1 PO III 535.

2 Cf. Edward O. Wilson, *Sociobiology, The New Synthesis*, (Cambridge, Mass., 1975), p. 13.

3 PO III 535.

the neurophysiological processes involved nor are we compelled to suppose the existence of 'sensations of innervation'. Even so there remains the undeniable impression of our consciousness that we ourselves are the initiators of certain so-called voluntary movements about which we are informed directly through corresponding kinæsthetic sensations as well as indirectly through the effects upon our sensory system following changes of its relative position. The vast majority of these changes are merely spatial while others may also involve qualitative and intensive alterations in the impression of the objects themselves.

> Now, is it possible for us to recognize the movements of our hands and eyes as spatial changes, without knowing this beforehand, and to distinguish them from other changes concerning the properties of things? Yes, I believe it is! It is an essentially distinctive feature of spatial relations that they are variable relations between objects, that do not depend on the quality and mass of the objects, while all other real relations between objects do depend upon the properties of the objects.[1]

Thus a certain subset of alterations among the impressions stands out as clearly identifiable. These are changes that affect only the local signs of any given impression. Such changes we simply *call* spatial. This is born out especially in the case of visual perception. Regardless of the content of any given impression on the retina, its accessory signs a', a", a"'... (which we gradually learn to interpret as *local* signs) may be altered by ocular movement into a new series of signs b', b", b"'..., and *vice versa*, while the remaining (qualitative and intensive) determinations of the original impression remain constant.

The above account, of course, does not explicitly specify which elements in the constitution of the general perception of space are given a posteriori and which, if any, must be given prior to any experience. In *The Facts in Perception*[2], Helmholtz's most comprehensive epistemological essay, his exposition of the various issues involved is much more thorough. All conscious data are alike in their merely psychological nature. In this

1 "Können wir nun die Bewegungen unserer Hände und Augen als Raumänderungen erkennen, ohne dies vorher zu wissen, und von anderen Aenderungen, welche die Eigenschaften der Dinge betreffen, unterscheiden? Ich glaube, ja! Es ist ein wesentlich unterscheidender Charakter der Raumbeziehungen, dass sie veränderliche Beziehungen zwischen den Substanzen sind, die nicht von deren Qualität und Masse abhängen, während alle anderen reellen Beziehungen zwischen den Dingen von deren Eigenschaften abhängen." (VR I 356).

2 'Die Tatsachen in der Wahrnehmung', (lecture delivered in Berlin, 1878), in SE.

fundamental respect our sensations should not be distinguishable from other psychological states. The reason that they are in fact distinguished from mere sentiments, say, is that psychic data are apperceived under radically different forms and accordingly classified as internal or external phenomena. The distinctive criterion which enables us to distinguish the outer sense and its external phenomena is precisely the one mentioned above: sensations belonging to certain modalities (viz. those which have reference to objects in space) turn out to be noticeably alterable by *voluntary motor impulses*, whereas psychic states such as memories, intentions, desires and the like, are not. Hence when we call the relations which are thus alterable *spatial*, then perceptions of *psychic* activities clearly do not enter into such relations at all. On the other hand, alteration upon voluntary movement being their sole distinctive mark, all sensations of the outer sense must occur under some kind of innervation, that is, they must be spatially determined. Accordingly, space can only appear to us as invested with the qualities of our kinæsthetic sensations, as something through which we look around and move about. In this sense, therefore, the perception of space is a *subjective* form of perception just as the sense qualities red, sweet or cold.[1]

The above analysis also entails that space is a *necessary* form of external perception since we classify as belonging to the external world precisely all that which we perceive as spatially determined in the defined sense. Thus, the proposition 'all external perception is spatial' is analytic since we *call* 'external' that which is perceived as being spatially determined in the manner described.[2]

Finally, space is an *a priori* form of perception[3] in so far as spatial perception depends on the possibility of voluntary motor impulses requiring mental and physical abilities with which the human organism must be endowed before spatial perception can occur.[4]

Clearly, Helmholtz's empirical theory is a far cry from Kant's original rationalist reconstruction of the perceptual process and its a priori foundations. The significance of the a priori is empirically redefined as the relative contribution of organic structures and of mediating mental

1 EW 124.

2 Ibid.

3 "[E]ine gegebene, vor aller Erfahrung mitgebrachte Form der Anschauung" ('Die Tatsachen in der Wahrnehmung', SE 117; henceforth the title of this important epistemological essay by Helmholtz will be abbreviated as TW).

4 EW 124.

activities to our knowledge of the world. Kant's doctrine of the synthetic a priori is hanging in the air. Space is a three-dimensional manifold of indefinite extension in which different entities coexist and within which magnitudes can be compared through the observation of point coincidences. These are the essential features of spatial perception all of which can be derived from observable consequences of controlled voluntary movements. Poincaré likewise regarded movement as the foundation of real space.[1] Space in this highly abstract sense is a purely formal schema without specific content.[2] As such, it is indeed a priori since its very formation is conditioned by the discovery of variability upon movement.

However, any more specific determinations like the axioms of Euclidean geometry cannot be regarded as necessary constituents of the formal schema. Geometry is a physical science, its axioms being applicable to the facts of experience only if they are tested and confirmed by actual measurements. Historically they may have been derived, intuitively, from certain recurrent primitive observations together with low-level interpretative theories. Even before the development of non-Euclidean geometries Newton already asserted in his Introduction to the *Principia* that geometry is dependent upon mechanical practice.[3]

If the axioms of geometry were a priori and necessary propositions of a 'transcendental geometry', they would confine the possible content of the spatial form of perception in accordance with certain specifiable laws, thus putting definite constraints upon the possible objects of perception. However,

> ...those very grounds that allow us to infer that the form of intuition of space is transcendental, do not yet necessarily

---

1 Henri Poincaré, *The Value of Science*, G.B. Halsted (tr.), (New York, 1907), p. 48. However, cf. Piaget's 'naturalistic' criticism of Poincaré's idea that the "shifting group", which he had rightly established at the development source of sensorimotor space, constitutes an *a priori* form of our activity and thought. As opposed to this view Piaget maintains that the shifting group *becomes* necessary by the gradual organization of actions. Cf. Jean Piaget, *Psychology and Epistemology, Towards a Theory of Knowledge*, Arnold Rosin (tr.), (Penguin Books, 1977), pp. 16-7. Clearly, Helmholtz's theory is closer in this respect to Piaget's developmental views than to Poincaré's more static position.

2 "ein inhaltsleeres Schema"; H. von Helmholtz, 'Über den Ursprung und die Bedeutung der geometrischen Axiome', (1870), SE 2.

3 "Fundatur igitur Geometria in praxi mechanica." (quoted in EW 162).

> suffice...to prove at the same time that the axioms too are of transcendental origin.[1]

In his celebrated popular essay 'The Origin and Meaning of Geometric Axioms'[2] Helmholtz dealt a severe blow to the Kantian doctrine of synthetic a priori propositions. Riemann, in the course of his meta-mathematical investigations of the foundations of analytic geometry had established that

> the essential foundation of any given system of geometry is the expression it gives for the distance between two points lying in any direction from one another, beginning with one which is infinitesimal.[3]

The most general form of this expression, taken from analytic geometry, is an equation according to which the distance of two infinitely near points equals the square root of a homogeneous second degree function of the differentials of their coordinates. Using this algebraic expression as an axiom Riemann deduced the conditions of free mobility of rigid bodies (or point systems). He called certain quantities yielded by the calculation—and coinciding with Gauss's measure of surface curvature when they are expressed for surfaces—the *measure of curvature* of the space at a particular point when they have the same value in all directions for that point. Riemann then demonstrated by means of the above axiom that free mobility of point systems is possible only in spaces of constant curvature.

Helmholtz, who had been working on very similar meta-mathematical problems before Riemann's doctoral dissertation (or *Habilitationsschrift*) had been published, reversed Riemann's reasoning. Starting from what he took to be a fact of observation[4], viz., that rigid bodies can freely move in space without change of form or size, he proceeded to demonstrate the necessity of the algebraic equation which in Riemann's presentation functioned as a mere hypothesis. Helmholtz's proof is presented in his 'The *Facts* underlying Geometry'[5], whose title marks the systematic distinction

---

1 "...diejenigen Gründe, welche schliessen lassen, dass die Anschauungsform des Raumes transzendental sei, genügen...noch nicht notwendig um gleichzeitig zu beweisen, dass auch die Axiome transzendentalen Ursprungs seien." (TW, in SE 122).

2 'Ueber den Ursprung und die Bedeutung der geometrischen Axiome' (1870).

3 H. von Helmholtz, 'The Origin and Meaning of Geometric Axioms (I)', (1870), in *Selected Writings of Hermann von Helmholtz*, Russell Kahl (ed.), (Middletown, Conn., 1971), p. 254; EW 13. The Kahl edition will be referred to below as 'Kahl'.

4 Cf. EW 15. Of course, this is a highly problematic assertion which will be discussed below.

5 SE 36-55; 'Ueber die *Tatsachen* die der Geometrie zugrunde liegen', (emphasis added), EW 39-58.

between his presentation and Riemann's *Habilitationsschrift* entitled 'The *Hypotheses* underlying Geometry'.[1] Riemann had first developed the general consequences of his initial hypothesis. Only then did he introduce as a further postulate that rigid point systems of finite magnitude be freely moveable in all directions without extension. This led him to the particular case of real space which satisfies that requirement. Helmholtz, however, raised a further foundational question. Might not Riemann's 'secondary' postulate, which distinguishes real space from other extended manifolds, have a substantial bearing upon the foundation of that basic proposition—which defines the concept of the length of a line segment in terms of a specific function of the differentials of the coordinates? Helmholtz showed that it did. Not only could this postulate of free mobility of rigid bodies provide a real foundation for Riemann's systematization of non-Euclidean geometries, but taking it as one's basic postulate from the very outset Riemann's initial hypothesis could be demonstrated to be deducible from much less restricted assumptions asserting nothing more specific than the monodromy of space and the congruence of any two of its parts under abstraction from their extension. Thus Helmholtz concludes:

> The fact that congruence is independent of place, of the direction of the spatial structures covering each other, and from the path along which they have been moved towards each other, that is the fact upon which the measurableness of space is based.[2]

It remains of course highly problematic, philosophically, when Helmholtz presents this fundamental fact as a fact of *observation*. For how do we know whether rigid bodies are 'really' freely movable and whether our measuring instruments are 'really' invariable in form? The concept of *congruence* being defined in terms of the coincidence of pairs of points of *rigid bodies*, it can of course, on pain of vicious circularity, no longer be utilized in defining what rigid bodies are. A conventionalist solution was proposed by most philosophers of the generation following Helmholtz. Helmholtz himself, however, seems to vacillate between a realist and a conventionalist interpretation. To be sure, he nowhere favors a radical conventionalist position. Yet, his essential insight into the theoretical interdependence of physics and

---

1 'Ueber die *Hypothesen*, welche der Geometrie zu Grunde liegen' (emphasis added).

2 "Die Unabhängigkeit der Kongruenz vom Orte, von der Richtung der sich deckenden Raumgebilde, und von dem Wege, auf dem sie zueinander geführt worden sind, ist die Tatsache, auf welche die Messbarkeit des Raumes basiert ist." [H. von Helmholtz, 'Über die Tatsachen die der Geometrie zugrunde liegen', (1868), SE 55].

geometry clearly adumbrates a pragmatic point of view entirely in line with his general pragmatic theory of truth.[1]

Thus the most general analytic formalization of our general notion of space as an extended three-dimensional manifold through which bodies can freely move and within which magnitudes can be compared does not entail any special characteristics of our own space as, possibly, expressed by the axioms of Euclidean geometry. These are not necessities of *thought*. Kant already realized this much. But neither are they necessities of *perception* since it follows from the above analysis that all spaces of constant curvature would be equally susceptible to the kind of observational experiments to which the human organism is predisposed and through which it learns to distinguish the local signs from the remaining determinations of the impressions it receives. Hence all such spaces are in principle equally well *perceivable*. Helmholtz even shows in minutely detailed analyses how certain non-Euclidean spaces would appear to us if they existed. He concludes:

> These remarks will suffice to show how we can infer from the known laws of our sensible perceptions the series of sensible impressions which a spherical or pseudo-spherical world would give us, if it existed.... [W]e can represent to ourselves the look of a pseudo-spherical world in all directions, just as we can develop the conception of it. Therefore it cannot be allowed that the axioms of our geometry depend on the native form of our perceptive faculty or are in any way connected with it.[2]

In fact, in the course of history the axioms must have been suggested by the daily recurrence of everyday experiences and by the intuitive apprehension of typical geometrical relations obtained by attentive observations,

> an intuition of the kind the artist possesses of the objects he is to represent and by means of which he decides surely and accurately whether a new combination which he tries corresponds to their nature. It is true that we have no word but *intuition* to mark this, but it is knowledge empirically gained by the aggregation and reinforcement of similar recurrent impressions in memory, not a transcendental form given before experience. That other such empirical intuitions of fixed typical relations, when not clearly comprehended, have

1 Cf. EW 24 f.

2 Kahl, 262; EW 23.

> frequently been taken by metaphysicians for a priori principles is a point on which I need not insist.[1]

The immediacy, the astounding clarity and profound penetration with which Helmholtz fathomed the revolutionary philosophical implications of the development of non-Euclidean geometries are nothing short of admirable. The idea of geometry as a physical science, its assertions being in principle refutable by observable facts; the idea of an axiom system of 'pure' geometry as an uninterpreted calculus without any real import unless conjoined with certain principles of mechanics (such as the axiom of inertia or the proposition that the mechanical and physical properties of bodies are independent of place); the idea—and its detailed empirical elaboration—of the profound and extremely complex origins of our perceptions of reality in spite of their apparent spontaneity and immediacy, as well as the consequent formulation of a pragmatic theory of truth; finally, the persistent appreciation of Kant as the critical epistemologist who provided the ultimate leverage against any future metaphysics and whose comprehensive analyses, rather than being weakened, paradoxically gained in strength by the refutation of some of his doctrines[2]; many of these and related Helmholtzian ideas would reverberate through the decades to come, spur research in metamathematics and philosophy of science, in cognitive philosophy and in epistemology, and lend momentum to a philosophical movement whose influence has not yet subsided. Yet, at the same time as these inspiring ideas merged into the mainstream of intellectual evolution, as they were 'purged' of impurities and as their rougher edges were carefully removed by the ensuing anti-naturalist framework of twentieth century philosophy, they became like polished pebbles at the bottom of a stream, no longer capable of stirring the water at the surface or of changing its course. Assimilated to their present environment they ceased to challenge it.

Helmholtz embodied the ideal of the *homo universalis*. His astounding accomplishments ranged over many fields of study. While his fruitful ideas developed together in one and the same mind, they stimulated each other reciprocally. But once disseminated, they were channeled into divergent directions, encapsulated by the 'appropriate' disciplines, and isolated from their original context. Helmholtz had been working, 'unorthodoxly', on both sides of the fence which later philosophers deemed necessary to erect between philosophy and empirical science. Thus various interesting ideas of his were

---

1 Kahl, 264; EW 25.

2 H. von Helmholtz, 'The Origin and Meaning of Geometric Axioms (II)', *Mind*, 3 (1878) 10, pp. 212-25; also in Kahl, cf. esp. pp. 361-2.

arbitrarily ruled out of court by subsequent generations as not worthy of philosophical consideration. This holds particularly true of his controversial theory of unconscious inferences whose systematic neglect by 'professional' philosophers and psychologists has been nothing short of detrimental to the development of epistemological inquiries slightly more comprehensive and inspiring than our present scholastic era has shown itself capable of producing.[1] The next chapter will be devoted to a discussion of the psychological hypotheses of Helmholtz's theory of knowledge.

1 Cf. p. 184, n.1.

X

HELMHOLTZ'S THEORY OF UNCONSCIOUS INFERENCES

1. The need of an empirical, non-introspective psychology.

As we have seen, the spatial determinations of the sensory input on the receptor organs is not strictly due to immediate sensations alone, i.e., their formation cannot be adequately explained by purely physiological processes. For all sensations, including the local signs, are merely empty symbols which our intellect must learn to interpret. Thus, for a sound and comprehensive theory of perception the physiologist must enter the field of psychology.[1]

However, it seems to Helmholtz that contemporary psychology is of no avail. The kind of psychic activities that must be postulated as entering into the process of perception is subliminal and largely beyond the control of the will. If this were not the case, that is, if we were directly conscious of the sensations instead of the perceptions, then we could easily sort out those elements of our conscious perceptions that are due to pure sensations and those that are the result of interpretation in the light of previous experience. The entire contro-

1 For Helmholtz's tripartite disciplinary division between the dioptrics of the eye, the neurophysiology of the visual system, and the psychology of perception, cf. VR 269. The very same distinctions were already made by Berkeley in his *The Theory of Vision Vindicated* (London, 1733), p. 43 (also in *The Works of George Berkeley, Bishop of Cloyne*, A.A. Luce and T.E. Jessop (eds.), (London, 1948-57) I, p. 266):

> "To explain how the mind or soul simply sees is one thing, and belongs to philosophy. To consider particles moving in certain lines, rays of light as refracted or reflected, or crossing, or including angles, is quite another thing, and appertaineth to geometry. To account for the mechanism of the eye is a third thing, which appertaineth to anatomy and experiments. These two latter speculations are of use in practice, to assist the defects and remedy the distempers of sight... But the former theory is that which makes us understand the true nature of vision, considered as a faculty of the soul."

versy between Helmholtz and Hering could never have arisen in the first place. However, many illustrations can be adduced, all of which indicate

> that we are exceedingly well trained in finding out by our sensations the objective nature of the objects around us, but that we are completely unskilled in observing the sensations *per se*; and that the practice of associating them with things outside of us actually prevents us from being distinctly conscious of the pure sensations.[1]

Thus, Helmholtz complains:

> It is hard to determine the nature of the mental processes which transform the sensation of light into a perception of the external world. Unfortunately, psychology is of no assistance, since up till now psychology has used introspection as the only method for obtaining knowledge, whereas in this case we are concerned with mental operations about which introspection is utterly silent and whose existence is to be inferred, rather, from physiological investigations of the organs of sense. Consequently, most psychologists have immediately classified the mental operations in question as sensory perception, with no attempt being made to obtain any further explanation concerning them.[2]

Kant, in overlooking the problem of localization, likewise comprised the complex processes of sensory perception into one single act while attributing to the mind a synthetic function determining no more than the *general form* of external perception which assimilates external sensory stimuli to intelligible structures. The interpenetration of mental and physiological functions in the formation of minutely detailed individual perceptions of objects in specific spatio-temporal arrangements he never made part of his philosophy. In fact, such a conception might be regarded as arising from a train of thought entirely alien to the Kantian system. For the psychic activities Helmholtz postulates are essentially *accommodating* rather than assimilat-

---

1 PO III 9.

2 "Die Natur der psychischen Prozessen zu bestimmen, welche die Lichtempfindung in eine Wahrnehmung der Aussenwelt verwandeln, ist eine schwere Aufgabe. Leider finden wir bei den Psychologen keine Hülfe, weil für die Psychologie die Selbstbeobachtung bisher der einzige Weg des Erkennens gewesen ist, wir es aber hier mit geistigen Thätigkeiten zu thun haben, von denen uns die Selbstbeobachtung gar keine Kunde gibt, deren Dasein wir vielmehr erst aus der physiologischen Untersuchung der Sinneswerkzeugen schliessen können. Die Psychologen haben daher die geistigen Acte, von denen hier die Rede ist, auch meist unmittelbar zur sinnlichen Wahrnehmung gerechnet, und keinen näheren Aufschluss über sie zu erhalten gesucht." (VR I 111).

ing functions. Kant's theory of the incognizable *Ding-an-sich*, however, as well as the idealist tendencies which generally characterize his critical philosophy are certainly not conducive to epistemological analyses along Helmholtzian lines.[1]

What, then, is the nature of these 'lower psychic activities' allegedly involved in perception? Without much scrutiny or theoretical scruples Helmholtz seems to accept as exhaustive the following alternation: either our perceptions are the direct results of sensation, or else some kind of *judgement* must be involved. And since the former alternative has been refuted the latter must hold. Helmholtz's logical characterization of the psychic activities may seem to be justified in view of the indirect, hence inferential, nature of the perceptual knowledge they yield. Similarly the astronomer, while studying the skies through his telescope, must *judge* what he perceives and must *compute* the positions of the stars in space, their relative distances etc., from the perspective images obtained of them at different times and from different parts of the earth's orbit. His conclusions, to be sure, are based on conscious knowledge of the laws of optics. In ordinary vision such knowledge is indeed lacking. Yet the perceptual knowledge acquired in mature vision must similarly be obtained through inferential acts, the only difference being that in the latter case such acts occur at an unconscious level.[2]

Helmholtz's theory of unconscious inferences and the implied analysis of cognition and of mental activities was scandalous not only to many contemporary psychologists but even more so to the subsequent generations of philosophers who were somehow related to the rationalist movement initiated or revived by Frege.[3] Thus, his stimulating psychologistic suggestions con-

---

1 Sometimes Helmholtz himself seems to be aware of this epistemological discrepancy (cf. VR II 338; also VR II 356, and TW, EW 143).

2 Cf. PO III 4.

3 That Frege's philosophy, and the ensuing tradition of analytic philosophy, is to be classified within the tradition of Western *rationalism* has been forcefully argued by H.D. Sluga, 'Frege and the Rise of Analytic Philosophy', *Inquiry*, 18 (1975), pp. 471-87; also id., *Gottlob Frege*, (London, 1980), esp. ch. II.

To interpret Frege's critique of psychologism as (somehow) linked to a basic philosophical critique directed against German *Idealism*, as Dummett has suggested [cf. his *Frege. Philosophy of Language*, (London, 1973)] is clearly to misconstrue the fundamental, strongly ***anti-naturalist*** inspiration that in fact animated Frege's attempt at philosophical reconstruction, and which subsequently became one of the defining characteristics of Anglo-American philosophical thought during the first half of the twentieth century. As I have indicated (cf. ch. VIII), intellectual life in mid-nineteenth century Germany was dominated not by German Idealism, but rather by various brands of

cerning the proto-intellectual foundations of human thought were neglected if not entirely ignored by professional philosophy.[1]

Perceptual judgments play a crucial role in the formation of normal perception. Most spatial perceptions (except perhaps the localization of direction, as

---

naturalism allowing only for philosophical positions that were scientifically informed. Helmholtz was one of the chief exponents of this philosophical movement.

1 Cf. M. Schlick's condescending comment:

> "On the celebrated theory of 'unconscious inferences'...we shall just briefly make the following comments. Modern psychology energetically rejects the concept of unconscious inference, because it rightly considers thought—the logical process—to be exclusively a function of *consciousness*. It may be asked whether Helmholtz merely uses an unsuitable terminology, or whether the improper terminology is also the expression of thoughts which do not stand up to rigorous epistemological criticism. We believe that Helmholtz's account, within broad limits, allows the first and favourable interpretation and therefore in fairness calls for it." (EW 176).

Again, for a professional *psychologist's* opinion representative of the general consensus of the 'scientific community' during the first half of the 20th century, cf. Boring's comment on Helmholtz's explanation of the phenomenon of simultaneous contrast: "In his [Helmholtz's] perplexity he resorted to the theory that the opposite color is seen as an illusion of judgement." This notion, Boring explains, Helmholtz generalized under the concept of unconscious inference. However (so Boring argues) "this is a negative explanation...essentially a confession of ignorance." [E.G. Boring (1942), pp. 167-8].

It is amusing (and redeeming) to compare these smug and seemingly irrevocable verdicts by leading scientists one or two generations ago to the bolder but nonetheless emphatic opinion of a highly influential contemporary theoretician in the area of epistemology and cognitive psychology which entirely reverses the 'common-sensical' consensus held before. Thus, R.L. Gregory places his program of research into the theory of perception squarely within the tradition instigated and inspired by Helmholtz's theory of unconscious inferences:

> "It is the fact that behavior does not need continuous, directly appropriate sensory data that forces upon us the notion of inference from available sensory and brain-stored data. This account is very much in the tradition of the nineteenth-century polymath physicist and physiologist Hermann von Helmholtz, who described perceptions as 'unconscious inferences'. This notion was unpalatable to later generations of psychologists, who were over-influenced by philosophers in their role—sometimes useful but in this case disastrous—of guardians of semantic inertia: objecting to inference without consciousness. But with further data on animal perception, and computers capable of inference, this essentially semantic inhibition has gone." (*Concepts and Mechanisms of Perception*, (New York, 1974), p. xx).

von Kries argues[1]) are acquired on the basis of numerous and varied experiences from early childhood onwards. We must learn to distinguish the local signs and their peculiar relations from the qualitative and intensive signs as received through outer sensing. Subsequently, the local signs must be interpreted specifically for the formation of a subjective spatial arrangement of the received impressions. The learning process is not merely a passive absorption of regular inputs. It involves active *experimentation*. For it is only through our ability to deliberately *change* the world and our relation to it that we come to know and understand it through the appearances it casts. Thus Helmholtz writes:

> In carrying out these movements and receiving the expected visual images we translate our representation, as it were, back again into the real world and check whether the translation corresponds with the original in order to convince ourselves by experiment about the correctness of our representation. I believe this latter point, in particular, deserves special consideration. The interpretation of our sensory impressions is founded upon experiment and not on mere observation of external events. Experiment teaches us that the connection between two events exists *at any arbitrary moment we happen to choose* under arbitrarily variable conditions in other respects.... Mere observation, no matter how often repeated under numerously varied circumstances, hardly ever guarantees us the same certainty of knowledge.[2]

In this crucial notion of experimentation Helmholtz draws a striking analogy between scientific and perceptual knowledge. While thus stressing the hypothetical and pragmatic nature of all our epistemic achievements, his 'activist' theory of knowledge also lays the foundation for a realist epistemology:

---

1 PO III 567 ff.

2 "Indem wir diese Bewegungen ausführen, und die erwarteten Gesichtsbilder erhalten, übersetzen wir unsere Vorstellung gleichsam wieder zurück in das Gebiet der realen Welt, und erproben, ob die Rückübersetzung mit den Originalen zusammenstimmt, um uns so durch das Experiment von der Richtigkeit unserer Vorstellung zu überzeugen. Ich glaube, dass namentlich dieser letztere Punkt wohl zu berücksichtigen ist. Die deutung unserer Sinnesempfindungen beruht auf dem Experiment und nicht auf blosser Beobachtung äusserer Geschehens. Das Experiment lehrt uns, dass die Verbindung zwischen zwei Vorgänge *in jedem von uns gewählten beliebigen Augenblicke* bestehe, unter übrigens von uns beliebig abgeänderten Verhältnissen.... Blosse Beobachtung gewährt uns kaum je dieselbe Sicherheit der Kenntniss, trotz noch so häufige Wiederholung unter vielfach veränderten Umständen." (VR I 355).

> The correspondence between visual perception and the external world is thus entirely, or at least mainly, founded on the very same ground upon which all our knowledge of the real word is founded, viz. upon experience and the continual experimental testing of its correctness as carried out with every movement of our body.[1]

This emphasis on action (rather than on pure observation or passive contemplation) as the sole source of true knowledge of the independent structures of the real world is even more clearly expressed in Helmholtz's 1892 lecture on Goethe entitled *Goethe's Anticipations of Later Scientific Theories*. Here he discusses Goethe's Faust as an epistemological allegory.[2] He sums up his interpretation as follows:

> It is only through action that Faust saves himself from the unsatisfactory state of introverted knowing and brooding, in which he cannot hope to arrive at the certain possession of truth and does not know how to comprehend reality... The epistemological counterpart of this scene is the fact that the efforts on the part of the various philosophical schools to justify the belief in the existence of reality was doomed to remain fruitless as long as they started from mere passive observation of the external world. They could not escape from their world of symbols [i.e., the arbitrary sign language of sensory information]; they did not recognize that man's voluntary actions constitute an indispensable part of our cognitive resources. We have seen that our sensory impressions are no more than a sign language informing us about the external world. Human beings must first learn to understand this symbolic system. And that happens when we observe the success of our actions and thus learn to discriminate between the changes in our sensory impressions that follow our intentional actions and those that occur independently of our will... Accordingly, an epistemology that builds upon the physiology of the senses has

---

1 "Die Uebereinstimmung zwischen den Gesichtswahrnehmungen und der Aussenwelt beruht also ganz oder wenigstens der Hauptsache nach auf demselben Grunde, auf dem alle unsere Kenntniss der wirklichen Welt beruht, nämlich auf der Erfahrung und der fortdauernden Prüfung ihrer Richtigkeit mittelst des Experiments, wie wir es bei jeder Bewegung unseres Körpers vollziehen." (VR I 365).

2 For brevity's sake I must leave the interesting details of the various metaphors Helmholtz uncovers untouched.

> to instruct people to move into action in order to obtain certain knowledge of reality.[1]

But we are running ahead of the story. Let me return to Helmholtz's theory of perceptual judgment and discuss it step by step.

Most of our present sensations are immediately joined and supplemented by relevant memory images and, according to Helmholtz, interpreted in terms of these. The apperception of a familiar room brightly lit by the sun is abundantly accompanied by vivid sensations. Later, in the evening twilight, and even at night in complete darkness, we can still move about in that same room and locate objects in it even though the images are utterly insufficient for recognizing any items without some previous acquaintance with them.

> Thus, by continually reducing the material that appeals to the senses, the perceptual image (*Anschauungsbild*) can ultimately be traced back to the pure memory image (*Vorstellungsbild*) and may gradually pass into it. In proportion as there is less and less material appeal to the senses, a person's movements will, of course, become more and more uncertain, and his apperception less and less accurate. Still there will be no peculiar abrupt transition, but sensation and memory will continually supplement each other, only in varying degrees.[2]

---

1 "Faust rettet sich aus dem unbefriedigten Zustande des in sich selbst gewendeten Wissens und Grübelns, wo er nicht zum sicheren Besitz der Wahrheit zu kommen hoffen darf und die Wirklichkeit nicht zu erfassen weiss, zur That.... Das erkenntniss-theoretische Gegenbild dieser Scene liegt nun darin, dass die Bemühungen der philosophischen Schule die Ueberzeugung von der Existenz der Wirklichkeit zu begründen, erfolglos bleiben mussten, so lange sie nur vom passiven Beobachten der Aussenwelt ausgingen. Sie kamen nicht heraus aus ihrer Welt von Gleichnissen; sie erkannten nicht, dass die durch den Willen gesetzten Handlungen des Menschen einen unentbehrlichen Theil unserer Erkenntnissquellen bildeten. Wir haben gesehen, unsere Sinneseindrücke sind nur eine Zeichensprache, die uns von der Aussenwelt berichtet. Wir Menschen müssen erst lernen, dieses Zeichensystem zu verstehen, und das geschieht, indem wir den Erfolg unserer Handlungen beobachten und dadurch unterscheiden lernen, welche Aenderungen in unseren Sinneseindrücken unseren Willensacten folgen, welche andere unabhängig vom Willen eintreten.... Auch die auf die Physiologie der Sinne gestützte Erkenntnislehre [muss] den Menschen anweisen, zur That zu schreiten, um der Wirklichkeit sicher zu werden." ('Goethe's Vorahnungen kommender naturwissenschaftlichen Ideen', VR II 359-60; also cf. PO III 30-1).

2 PO III 11.

Similarly, a white sheet of paper in the dark is perceived as white even though it is less luminous than a grey sheet of paper in bright sunlight.[1] Finally, perceptual illusions must be explained as arising from mistaken judgment rather than from 'errors' in the afferent nervous system. The illusion of stereoscopic luster is mediated through association with the standard perception of the gloss of a surface, which, although apparently a simple effect, is due to differences of coloring or brightness in the two retinal images of that surface.

Again, it is well known that whenever the cutaneous nerves are stimulated, such stimulations are always perceived as occurring in the corresponding peripheral surface of the skin even when they affect only the stem of the nerve center itself. Since the stimulation of the tactile nerves in the overwhelming majority of cases is due to influences that affect the terminal fibers of these nerves in the surface of the skin the judgment is inevitably led to the inductive inference that all such stimulations must be due to similar external conditions.[2]

An analogous example—and a favorite one with Helmholtz—is the following. Whenever the temporal part of the eyeball is mechanically stimulated (by exerting pressure upon it), we get the impression of a luminous object somewhere in the direction of the nasal bridge. This phenomenon is known as *pressure phosphene*. Stimulations of the optic nerve that are not produced by external light can be regarded as rare exceptions since the retina in the fundus of the firm eyeball is almost completely protected from the actions of all other stimuli. As the temporal parts of the retinas under ordinary conditions of vision are stimulated only by light rays coming from the

1 Referring occasionally to the imagination and very frequently to the attention Helmholtz formulates at least three general rules determining visual perception, or even perception in general. The first proposition states that we are easily and precisely aware only of those sensations which contain useful information concerning external objects while everything else (e.g., the blind retinal spot) goes by unnoticed. (Cf. VR I 104, 109 f., 291 f.; PO III 6).

A second rule stipulates that nothing in our perceptions can be recognized as due to sensation that can be altered, or even reversed, by factors demonstrably due to experience. (Cf. VR I 353; PO III 13).

Finally it is stated that "such objects are always imagined as being present in the field of vision as would have to be there in order to produce the same impression on the nervous mechanism, the eyes being used under ordinary normal conditions." (PO III 2; 14 f., 534 f.; cf. VR II 357; for illustrations of these various rules, cf. text below).

2 PO III 3. The phenomena of phantom limbs and other similar phenomena were already discussed and explained by Malebranche (RV, p. 25).

opposite direction, the perceptual judgment is led to an inductive inference in accordance with

> the general rule determining the ideas of vision that are formed whenever an impression is made on the eye:...such objects are always imagined as being present in the field of vision as would have to be there in order to produce the same impression on the nervous mechanism, the eyes being used under ordinary normal conditions.[1]

Mutatis mutandis, this rule expresses, of course, a general characteristic of all sense perceptions.

In all these cases of perceptual judgments the structure of the inference is essentially inductive. Numerous uniform experiences are accumulated and stored in the memory in the form of some implicit rule whose specific instances have long since been forgotten and which determines the interpretation of any new relevant sensation. Like causes are inferred from like effects under like circumstances. These unconscious inferences could be expressed in the form of syllogisms whose major premiss is inductively obtained. In the last example the corresponding syllogism might run as follows (Helmholtz does not spell this out):

(1) Every temporal excitation of the retina corresponds to a light ray passing through the median plane.

(2) A retinal place on the temporal side is now being stimulated.

(3) Hence I see (there exists) something luminous in the direction of the median plane.

Clearly, only the conclusion of this syllogism expresses a conscious 'sense datum'. If such sense data are complex, appearances to the contrary notwithstanding, then the underlying inference must occur subliminally, i.e., below the threshold of conscious thought. The same holds with respect to the peculiarities of spatial perception. They are subconsciously derived from the observed continuous variability of the local signs upon bodily or ocular movement. The spatial conception of the persistent coexistence of different things simultaneously together, and the related conceptions of object permanence (objects continue to exist unperceived) and of object constancy throughout varying perspective appearances, all require associative activities which are essentially inductive. It follows that such conceptions can be developed only gradually through experience, training and experiment. Piaget's later findings on the basis of extensive experimental research into the area of cognitive psychology—research in which Helmholtz himself never actually engaged—seem to bear out many of Helmholtz's

1 PO III 2.

theoretical suggestions. Indeed, Helmholtz often refers to the child's learning practices and its deliberate experiments in play and imitation when he discusses his empirical theory of perception.[1]

2. Helmholtz's theory not a mechanistic theory, but a truly cognitive theory of information processing.

But why does Helmholtz characterize these lower psychic activities, which mediate between the pure sensations and the conscious perceptions, as processes of *thought*? Helmholtz has stubbornly defended this controversial contention:

> ...the more attentively I have studied the phenomena, the more I have been impressed by the uniformity and harmony everywhere of the interplay of the psychic processes.... And so I have had no scruples in connecting and unifying the facts...by explanations which were founded essentially on the simpler psychic processes of the association of ideas.... The fundamental thesis of the empirical theory is: The sensations of the senses are signs for our consciousness, it being left to our *intelligence* to learn how to comprehend their meaning.[2]

The perceptual judgments are of a peculiar kind. They are unconscious ones, incapable of being expressed in words. They cannot be raised to the level of our natural consciousness which can perceive its own activities, reflect upon them and control them. They are associative processes that occur fast and imperceptibly in the dark background of our memory. It is not our conscious selves which draw these inferences but rather the gradually developing conceptions and ideas *in us*. In these perceptual 'judgments' ideas and sensory images take the place of words in ordinary judgments, and it is precisely because we can exert little or no influence upon them that their results seem to be forced upon us as though by some external agency. This explains why most perceptual illusions persist in spite of our improved insight into their mechanism. The underlying inferences tend to become conditioned responses, so thoroughly inculcated into the mind's habitual response patterns and so

1 SE 118 ff.; PO III 31.
2 PO III 533; emphasis added.

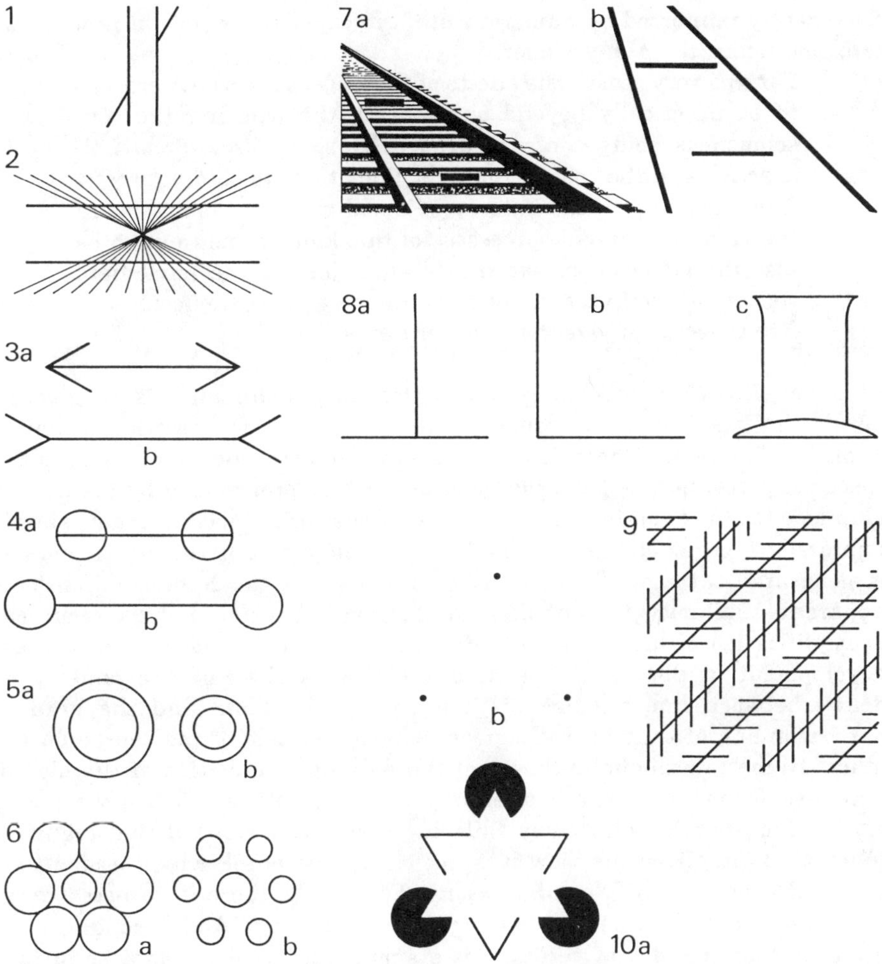

Fig. 21 — Perceptual Illusions. *(1) Poggendorff (1860): straight line; (2) Hering (1861): parallel lines; (3) Müller-Lyer (1889): lines of equal length; (4) Delboeuf (1892): lines of equal length; (5) Circles of equal diameter; (6) Titchener (1898): central circles of equal diameter; (7) Ponzo illusion: bars of equal length; (8) Vertical and horizontal lines of equal length; (9) Zöllner illusion: parallel lines; (10) Kanisza: (a) subjective contours and enhanced brightness (the effect can be enhanced by placing tracing paper over the display and/or by viewing it from a distance; (b) similar triangle with inferred boundaries but no subjective contours and no illusory brightness.*

consistently reinforced by numerous uniform experiences that in practice we cannot unlearn them any longer.

> For this very reason the effects of these ideas are so powerful as to be practically beyond our control, the will and the consciousness being confronted as if by some *force of nature*, exactly as in the case of the sensations that we obtain directly from outside. Thus, whatever is joined with the sensations in the results of psychic processes of this kind seems to us to be also the effect of an external agency, just as the immediate sensation itself is, and not something discovered by conscious free reflection, *thought out* by our own selves.[1]

It is not always clear, however, whether Helmholtz intends to give an intellectualistic account of sensitive functions or a mechanistic account of cognitive functions. There is an apparent similarity between Helmholtz's theory of perception and the theory of association propounded by the British empiricists to explain how complex ideas are derived from simple impressions, or, as Hume explained it, how the idea of necessary connection is generated through the associative processes of the human mind. This apparent similarity is contradicted, however, by Helmholtz's recurrent exemplification of his theory by referring to the perceptual and conceptual development in the child's learning-to-see, as well as by the analogy he draws between language acquisition on the one hand and the gradual understanding of the naturally given symbolic system of the senses on the other. Hume's mechanistic theory of the habitual acquisition of the idea of necessity, for instance, surely offers a plausible genetic account of the *source* of this idea but it notoriously fails to ascribe any constructive cognitive functions to it. Given his view it is not clear, first of all, why the idea thus acquired should be an idea of *necessity*. For, really, there is no necessity in the mind either but only a 'feeling of inevitability'. And secondly, even granting that the idea of necessity is a simple idea which cannot be further analyzed, it would still make sense for us to ask what difference the having or not having of this idea would make to intelligent creatures. Hume is silent on this point. Wittgenstein seems to raise the latter question. And again, this question is not a further request for meaning analysis, but rather a request to spell out the *cognitive functions* of the idea of necessity and of the psychic operations responsible for its emergence.[2]

---

1 PO III 541.

2 Cf. Barry Stroud, *Hume*, (London, 1977), pp. 68-95.

By contrast, Helmholtz's characterization of the pervasive inductive inferences unconsciously operative in human perception as elementary *thought* processes attributes from the very outset cognitive import to what Hume called the 'principles of human nature' as well as to the perceptions (or ideas) to which they give rise. And this means that the gap between simple sensations and elementary cognitive or epistemic facts is much more easily bridged on Helmholtz's account than on Hume's. The mind is already operative long before we become conscious of the results of its operations and can express them in words. The sensing of sense contents does not require a sudden and epistemologically puzzling transition from a mere physiological state to its epistemic assessment. The associative habits of the memory and our primitive ideas of space, of time, and of particular spatiotemporal objects have already developed through subliminal interactions between elementary mental activities and aggregates of regular impressions long before they are further shaped by linguistic expression.

3. Helmholtz's theory of a continuum of cognitive functions beyond the edge of consciousness and beyond the grasp of verbal articulation.

Helmholtz has become more cautious over the years regarding his theory of unconscious inferences. When it was first proposed (*Ueber das Sehen der Menschen*, 1855) he appealed to Kant's doctrine of the aprioricity of the idea of causality in an attempt to show that the unconscious interpretative activities of the mind are not merely mechanical processes but involve acts that are genuinely inferential:

> Should we conclude, then, that what I have called the thinking and inferring of representations isn't really thinking and inferring, but nothing more than a mechanically conditioned combination of ideas? I beg you to take one last further step with me, a step which will bring us back to where we began, to Kant. In order that a connection be brought about between the representation of a body of a particular appearance and in a particular situation and our sensations we must surely already have the representation of such bodies.... But how did we ever for the first time achieve the transition from the world

> of our nerve impressions to the real world? Apparently only by means of an inference.[1]

Thus what we find here is an indubitably mental, or cognitive, activity with synthetic functions, yet taking place unconsciously. However, Helmholtz's reasoning fails to carry conviction. For unlike the unconscious inductive inferences Helmholtz has postulated so far, the primordial causal inference which is responsible for our belief in an external world is *not* mediated through association but, as the Kantian Helmholtz of this early period expressly points out, is prior to *all* experience.

> Thus the investigation of sensory perception also leads us to the insight already found by Kant: that the proposition *'No effect without a cause'* is a law of our thought prior to all experience.[2]

However, in his *Treatise on Physiological Optics* and its extract *Recent Advances in the Theory of Vision* (1868) Helmholtz offers a much broader analysis of cognition in order to show that conscious thought constitutes no more than an extremely thin layer of the complex and highly stratified phenomena of mental life. He defends his thesis by pointing out that there exists a vast array of very precise cognitions which are largely protolinguistic, and, closely connected with this fact, that there are numerous functions of the intellect at lower, and sometimes even relatively high, levels of cognitive performance, which are proto-intellectual, that is, where cognitive operations, undoubtedly inferential in character, are nevertheless unconsciously carried out.

His logical theory, meanwhile, has undergone a major influence from J.S. Mill's *System of Logic*. He draws a contrast between logical and

---

1 "Somit wäre das, was ich früher das Denken und Schliessen der Vorstellungen genannt habe, nun doch wohl kein Denken und Schliessen, sondern nichts als eine mechanisch eingeübte Ideenverbindung? Ich bitte Sie, noch einen letzten Schritt weiter mit mir zu machen, einen Schritt, der uns wieder auf unseren Anfang, auf Kant, zurückführen wird. Wenn eine Verbindung zwischen der Vorstellung eines Körpers von gewissen Aussehen und gewisser Lage und unseren Sinnesempfindungen entstehen soll, so müssen wir doch erst die Vorstellung von solchen Körpern haben.... Auf welche Weise sind wir denn nun zuerst aus der Welt der Empfindungen unserer Nerven hinübergelangt in die Welt der Wirklichkeit? Offenbar nur durch einen Schluss." (VR I 115-6).

2 "Also führt uns die Untersuchung der Sinneswahrnehmungen auch noch zu der schon von Kant gefundenen Erkenntnis: dass der Satz: "Keine Wirkung ohne Ursache", eine vor aller Erfahrung gegebenes Gesetz unseres Denkens sei." (VR I 116). The use of the correlative terms *'Wirkung'* and *'Ursache'* is awkward, of course. What is meant is better stated as 'no event without a cause'.

'inductive' inferences. The former ("the inferences of the logicians"[1]) are syllogisms whose major premises are inductively obtained.

> The distinction between the logicians' inferences and the inductive inferences whose results show up in the perceptions of the external world as derived from our sense-impressions seems to me...to be inessential. For it consists mainly in this that the former is capable of being expressed in words while the latter is not, since in this case instead of words only sensations and memory images of sensations play a role. This very fact that they cannot be described in words, also makes it so very hard to talk about this entire domain of mental operations in the first place.[2]

Helmholtz's reasoning seems to be somewhat as follows. The fact that many 'inductive' inferences cannot be transformed into a precisely formulated syllogism militates against their being inferences only if their elements (present sensations, memory images, subliminal experiences, etc.) must be denied all epistemic status on the sole ground that they are not expressible in words nor elements of our natural consciousness. But by the same token a large segment of what we would ordinarily classify as cognitions or as having cognitive import would go by the board as well.

We must distinguish between '*Wissen*' (i.e., propositional knowledge) on the one hand, and '*Kennen*' (or knowledge-by-acquaintance, in the sense of the French '*savoir*') on the other. The latter involves non-verbal kinds of cognition.

> In addition to the kind of knowledge that works with concepts and therefore is capable of being expressed in words, there exists yet another area of representational competence, which combines only sensory impressions that cannot immediately be expressed in words. In German we call it '*das Kennen*'.[3]

---

1 "die Schlüsse der Logiker".

2 "Der Unterschied zwischen den Schlüssen der Logiker und den Inductionsschlüssen, deren Resultat in den durch die Sinnesempfindungen gewonnenen Anschauungen der Aussenwelt zu Tage kommt, scheint mir...nur ein aüsserlicher zu sein, und hauptsächlich darin zu bestehen, dass erstere des Ausdrucks in Worten fähig sind, letztere nicht, weil bei ihnen statt der Worte nur die Empfindungen und Erinnerungsbilder der Empfindungen eintreten. Eben darin, dass diese sich nicht in Worten beschreiben lassen, liegt auch die grosse Schwierigkeit, von diesem ganzen Gebiete der Geistesoperationen überhaupt zu reden." (VR I 358).

3 "Neben dem Wissen, welches mit Begriffen arbeitet, und deshalb des Ausdrucks in Worten fähig ist, besteht noch ein anderes Gebiet der Vorstellungsfähigkeit, welches

Thus we know a man, a road, an item of food and a fragrance without being able to give precise descriptions. This kind of knowledge-by-acquaintance constitutes cognition just as much as recognition does which presupposes it. It displays the highest degree of determinacy and certainty and is not inferior in these respects to the achievements of propositional knowledge (*das Wissen*). Helmholtz doubts whether among the ideas of the adult there are any cognitions (*Kenntnisse*) of this kind that require for their origin a source other than the unconscious activity of the memory.[1] Thus, the perceptions of our natural consciousness also differ from conscious logical thought by their immediacy and urgency, presenting themselves as it were spontaneously, "without conscious reflection"[2] and not subject to voluntary control.

Another important aspect of these non-verbal cognitions is the *practical* knowledge of how to innervate the motor nerves for perceptual purposes.[3] Ocular movement and adjustment have to be learned just as we must learn how to walk. Such apparently simple skills as how to focus the eyes on any given luminous object the newborn infant has yet to master. But once mastered, they become so thoroughly ingrained into our nature that their influence on perception is no longer noticeable. The acquired practical knowledge is stored in the form of sensory images and habits of the memory. Verbally, but very imprecisely, expressed while already resorting to the technical vocabulary of the physiologist its general form might read as follows: whenever I apply such precisely known voluntary impulses to the motor nerves of the two eyes I will be focusing on a point at 2' distance (that is, the point at the *fovea centralis* of each eye will be the image of a point at a distance of 2 ft.). This knowledge can function as a major premise in an unconscious inference whose minor asserts the kinæsthetic feeling that I am at present looking as stated in the major and whose conclusion asserts the perception of an object at 2' distance. Even when I know that the conclusion is false (e.g., when I am looking through a telescope) the illusory perception does not vanish. It presents itself as 'naturally given' though in fact it is not.

> We cannot eliminate the sensation of this illusion, we cannot eradicate the memory of its normal meaning, not even when we know that in the present case it doesn't apply; no more than we are able to erase from our minds the meaning of a word of

---

nur sinnliche Eindrücke combinirt, die des unmittelbaren Ausdrucks durch Worte nicht fähig sind. Wir nennen es im Deutschen das *Kennen*." (Ibid).

1 WA III 553; VR II 341; cf. L. Koenigsberger (1906), p. 428.

2 PO III 25.

3 "Diese Art des Kennens nennen wir ein *Können* (im Sinne des franzözische *savoir*) oder auch wohl ein *Verstehen* (zum Beispiel: ich verstehe zu reiten)." (VR I 359).

> our mother tongue, when it is once used for an entirely different purpose as a sign or a password.[1]

Helmholtz's theory of unconscious inferences now begins to take shape. Influenced particularly by Mill, he holds that the natural and original form of all human reasoning is induction by analogy. We naturally reason from particulars to particulars, sharing this faculty with other forms of animal life. The essential office of such reasoning is to amplify our stock of knowledge and to utilize it beyond what is presently known. This function can be performed—and is performed with, on the whole, a remarkable degree of precision and reliability—regardless of whether the various steps in such reasonings lend themselves to linguistic articulation or not. The former possibility serves the completely different purposes of communication and cultural transmission, of testing, scrutinizing and purifying our expanding stock of knowledge by the powerful means of conscious thought. But as far as the ampliative function of inductive connections is concerned, it is entirely immaterial whether these connections are mediated through terms or templates, through concepts or percepts.

> It is...clear that with such sensory memory images instead of words the same kind of connection can be constructed which, when expressed in words we would call a proposition or a judgment.[2]

Thus, it is in view of their *results* that "the unconscious processes of association of ideas going on in the dark background of our memory"[3] are to be identified as unconscious *inferences*:

> ...while in these cases no particular conscious inference may be present, yet the essential and original office of such an inference has been performed, and the result of it has been attained...These inductive inferences leading to the formation of our sense perceptions certainly do lack the purifying and scrutinizing work of conscious thinking. Nevertheless, in my

---

1 "Wir können die Empfindung dieser Täuschung nicht fortschaffen, wir können die Erinnerung an ihre normale Bedeutung nicht vertilgen, selbst wenn wir wissen, dass diese in dem vorliegenden Falle nicht zutrifft; ebenso wenig, als wir die Bedeutung eines Wortes unserer Muttersprache uns aus dem Sinne schlagen können, wenn es einmal als Zeichen oder Stichwort zu einem ganz anderen Zwecke angewendet wird." (VR I 361).

2 "Es ist...klar, dass man mit dergleichen sinnlichen Erinnerungsbildern statt der Worte dieselbe Art der Verbindung herstellen kann, die man, wenn sie in Worten ausgedrückt wäre, einer Satz oder ein Urtheil nennen würde." (VR I 360).

3 PO III 26.

> opinion, by their peculiar nature they may be classified as *inferences*, inductive inferences unconsciously formed.[1]

When Helmholtz further illustrates the pervasiveness of this kind of psychological rather than logical inferences by their natural preponderance in the '*Geisteswissenschaften*' (the humanities and the social sciences), his judgment is of course influenced by the predominantly historical and hermeneutical character of these sciences in the nineteenth century.

> The natural sciences are mostly capable of articulating their inductions in terms of explicitly formulated general rules and laws; the humanities, by contrast, are mainly dealing with judgments that depend on judiciousness and psychological discretion.[2]

Accordingly, Helmholtz distinguishes the inductive inferences of the humanities as 'artistic' inductions from the logical inductions of the hard sciences. For it is the artist rather than the scientist who can find his way through the bewildering complexities of human life, sift them and represent them with unexpected rhyme and rhythm, under a coherent and compelling perspective that does not rest on generally valid rules but on intuited typologies.

> Now this latter kind of induction, even though it cannot be reduced to the perfected form of logical inference nor to the level of articulate exceptionless laws, plays an immensely extensive role in human life.[3]

To deny all intelligent thought activity to these subliminal inductive operations, to deny them all rational or inferential status in spite of their

---

1 PO III 26-7. The Southall translation mistakenly translates the German '*Schluss*' (as in '*Inductionsschlüsse*' or '*unbewusste Schlüsse*') as 'conclusion' rather than as 'inference'. However, if '*Schlüsse*' was meant in the sense of 'conclusions', they could not be said to "lead to the formation of...sense-perceptions" (cf. passage quoted) because, if they are assumed to be the (conscious) *results* of (unconscious or subconscious) processes, they would then be identical with these sense-perceptions.

2 "Die Naturwissenschaften sind meist im Stande, ihre Inductionen bis zu scharf ausgesprochenen allgemeinen Regeln und Gesetzen durchzuführen; die Geisteswissenschaften dagegen haben es überwiegend mit Urtheilen nach psychologischen Tactgefühl zu thun." (VR I 172).

3 "Diese letztere Art der Induction nun, welche nicht bis zur vollendeten Form des logischen Schliessens, nicht zur Aufstellung ausnahmslos geltender Gesetze durchgeführt werden kann, spielt im menschlichen Leben eine ungeheuer ausgebreitete Rolle." (VR I 171).

being based on experience and guided by "experiments"[1] is to rob all conscious thought of its protolinguistic foundations and to render most of our cognitions, esp. our perceptual and psychological beliefs, miraculously instinctive and neuro-mechanically predetermined while leaving their remarkable adaptedness and efficiency virtually inexplicable. For against such a complete reduction of the inductive processes to fixed neuronal activities the same arguments would hold that refuted the intuitionists' theory of perception—it would leave out of account their conjectural and essentially accommodating function as well as the important role learning and experience play—and continue to play—in their formation and gradual perfection in the direction of an optimal fit between stimulus and response.

Or to put it another way, it is true that Helmholtz makes a clear distinction between psychological and purely physiological processes. However, by drawing this distinction he meant no more than to emphasize the fact that there exists a more or less well-defined set of physiological processes which are thoroughly studied and reasonably well known but which do not suffice to account for perception as a whole. Another set of factors has to be postulated which exhibit relations and potentials different from those of the former class of processes while bearing some distinct similarity to the activities of mental life with which we are familiar from everyday life. To classify the latter as psychological factors does not entail an implicit assertion of psychophysical dualism. On the contrary. To suppose a physiological foundation of experience and of the processes of learning and habituation would not in any way conflict with Helmholtz's overall analysis. Helmholtz himself suggests:

> If anyone objects to including these processes of association and the natural flow of ideas among the psychic activities, I will not quarrel over names.[2]

He allows a similar interpretation when, in arguing against the physiological explanation of the rivalry between the visual globes of the two eyes (which occurs whenever two conflicting images cannot be fused into one intelligible impression), he attributes the resulting phenomenon of the constant alternation of the two impressions to the variable disposition of the *attention* (a psychological rather than a physiological faculty) and adds:

> ...this alternation does not depend on some organic mechanism of the nervous system...at least on nothing more than underlies our mental activities.[3]

---

1 VR I 354; cf. (in Piagetian terms) the child's tertiary circular reactions towards the end of the sensori-motor period.

2 PO III 541.

3 PO III 500.

Thus, whatever the proper (presumably neurophysiological) explanation may be of the psychic activities operative in our perceptions (and in other cognitions inductively obtained), they undoubtedly constitute a set of factors which in view of their complexity, their flexibility and their capacity for development must be distinguished from the dioptrical and physiological determinants of mere sensations. This distinction remains valid no matter how far the materialist research program embodied in the nineteenth century scientific ideals may eventually succeed in closing the gap between growth and development.

4. Helmholtz's theory dogmatically dismissed by the twentieth century ban on psychologism. Yet his cognitive theory superior as compared to traditional alternatives.

It is clear that Helmholtz's basic point of view remains a physiological one. The distinction between sensation and perception (already to be found in Müller) and the assumption of the priority of the former leads him to a theory of perception radically different from the exclusively psychological sense data theories which have either simply identified the sensations with the perceptions or regarded the sensations as constitutive elements from the very outset of any given perception and the latter as the immediately given.

It is also clear that this physiological point of departure provided the impetus for a theory of mind that was increasingly non-rationalistic and evolutionary. The roots of rational thought extend deep into the subconscious layers of primitive intelligence.

> Inductive reasoning is the result of an unconscious and involuntary activity of the memory.[1]

Moreover, this psychic process is operative in man and animal alike and from the lowest to the highest levels of mental activity. The merely formal role Kant had attributed to reason has been extended by Helmholtz in an altogether unorthodox fashion: reason, involuntarily, contributes even to the *content* of our perceptions. The analysis of the concept of perception into elementary processes of *thought* seems to Helmholtz, *pace* Kant, the most essential advance of modern science.

---

1 PO III 28.

> The very first elementary perceptions already contain cognitive processes and proceed according to the laws of thought. Everything in perception that is added to the raw material of sensation, can be resolved in thought.[1]

Helmholtz's interesting theoretical suggestions concerning the 'ratiomorphic' activities of the mind opened up new avenues of research in functional and cognitive psychology which only very slowly came to arouse more widespread interest, J.M. Baldwin being one of the great pioneers in such studies. The results of these studies, however, were entirely ignored by the analytic philosophers and the logical empiricists of the first half of the twentieth century. Swayed as they were by the Cartesian ideas and ideals inherited from the rationalist, strongly anti-naturalist revival towards the end of the nineteenth century (cf. the Neo-Kantians and their strictly logical interpretation of Kant; von Brentano's intentionalist conception of consciousness with the world as its intentional correlate; Husserl's transcendental phenomenology; Frege's logicism etc.) they rejected the psychologism and biologism inherent in various 19th century research programs (e.g., those of Helmholtz, Spencer, Bain, Baldwin, i.a.) as flying in the face of reason which they viewed as an ultimate and autonomous authority capable of discerning eternal logical relations which do not depend on the so-called 'laws of thought' and their contingent genesis. To lay bare the 'viscera' of pure logical thought and to connect the sublime power of reason with the lower psychic activities to be found in man and animal alike struck them as laughable if not as perversely provocative. It would open the gates to cognitive relativism. All rational law and order would be gone.

Descartes had *defined* thought as that mental activity or state of which we are immediately conscious:

> *Thought* is a word that covers everything that exists in us in such a way that we are immediately conscious of it.[2]

He even deliberately extended the received usages of '*cogitare*' and '*penser*'. Thus he goes on to say:

> ...all the operations of the will, intellect, imagination and of the senses are thoughts.[3]

---

1 "Schon die ersten elementaren Wahrnehmungen enthalten in sich ein Denken und gehen nach den Gesetzen des Denkens vor sich. Alles was in der Anschauung zu dem rohen Materiale der Empfindung hinzukommt, kann in Denken aufgelöst werden." (SE 130; VR II 240; cf. WA III 553).

2 HR II 52.

3 Ibid.

His main thesis, then, was "that the essence of a mind is consciousness, or to be conscious".[1] This is also the fundamental premise presupposed by the method of introspective psychology. Of course, this thesis did not go unchallenged, not even in the rationalist tradition. While Leibniz agreed with Descartes, and against Locke, that the mind is always active, he denied that all its activities are conscious. In fact, no particular monad-of-monad perceptions are ever conscious. They are "*petites perceptions*" which are individually below the threshold of consciousness. It is only collectively that they result in the aggregate phenomenon of conscious molar experiences. Thus, man's conscious perception is always "confused".

Leibniz's theory of subliminal perception was an important innovation in psychological theory. Yet it did not bear much fruit. And this is not surprising. For rather than being a bold psychological conjecture with a well-defined research program and testable hypotheses about thought and perception it remained no more than a mere consequence of the systematic exigencies of his perplexing monadology. And this may have obscured the possibility of the *independent* significance of his doctrine of the unconscious.

Nevertheless, Leibniz's theory did help to prepare the way for the development of further psychological theories of the unconscious.[2] And there are striking similarities between Leibniz and Helmholtz on this account. For the unconscious, with Helmholtz, is not the negation of all consciousness but rather the 'confused' perception of general rules and validities which are the aggregate effects of numerous individual experiences whose specific characteristics are completely blurred in memory.[3]

I have already touched upon some of the similarities and dissimilarities between Helmholtz and Hume. To be sure, Hume's treatment of the various questions under consideration is philosophically more profound and systematic than Helmholtz's who enters the field of epistemology from the part of physiology and is therefore empirically better documented than Hume. Yet, in allowing for *gradations* of rational activity down to the level of 'merely' habitual inductions Helmholtz seems to improve upon Hume's naturalism in offering a more unified comprehensive theory of knowledge. Reason and habit are no longer treated as absolute contrasts as in Hume, the one more or less free, the other a blind psychological reflex in which 'nature' fortunately gets the better of our futile reason. Rather, habit itself is seen as

---

1 S.R. Schiffer, 'Descartes on his Essence', *Phil. Rev.* 85, 1 (Jan., 1976), p. 2.

2 Cf. L.H. Whyte, *The Unconscious Before Freud*, (New York, 1960).

3 Cf. VR II 172-3, 233.

exhibiting selective and adaptive patterns which demand that it be more broadly defined as including sub-rational processes. As Goodman has pointed out, not all regularities automatically generate inductive responses, while in other cases, as Mill has indicated, no regularity but a single instance is sufficient for us to confide in a universal generalization. Similarly, Helmholtz emphasizes that it is through habit that the individual perfects himself during his lifetime and that none of the acquired inferential habits—except the most primitive ones and those most widely confirmed and most deeply entrenched—are in principle unalterable or no longer adaptable to novel and unusual circumstances. Thus the function of consciousness, in securing new accommodations, remains available. This twofold psychological function of the mind's unconsciously judging what it perceives on the basis of previous discoveries, that is, of the spontaneous *habituation* of successful *accommodations*, adds a cognitive rational dimension to the conditioned but highly selective operations of the mind which is overlooked in Hume's merely mechanical 'repetition theory' of habit as "nothing but one of the principles of nature"[1] and of Nature as determining us "by an absolute and uncontroulable necessity...to judge as well as to breathe and feel".[2]

In Helmholtz's view the psychic activities operative as inductive reasonings in perception and in 'artistic induction' are undeniably ratiomorphic.

> These investigations...bring us face to face with a domain of mental activities which has been rarely mentioned in scientific inquiries so far, because it is so difficult to verbalize them at all. Indeed, they have been studied more commonly only in æsthetics, where they play a significant role under such obscure designations as "intuition", "implicit understanding", or "sensory reason". They are very incorrectly prejudged as proceeding in a vague, indeterminate, merely halfconscious manner, and that, as a kind of purely mechanical operation, they are subordinate to conscious thought expressible in language. I do not believe that the former and the latter can be distinguished in respect of the nature of the activity itself.[3]

---

1 David Hume, *A Treatise of Human Nature*, L.A. Selby-Bigge (ed.), (Oxford, 1888), p. 179.

2 Ibid., p. 183.

3 "Wir werden...durch diese Untersuchungen zu einem Gebiet von psychischen Thätigkeiten geführt, von denen bisher in wissenschaftlichen Untersuchungen wenig die Rede gewesen ist, weil es schwer hält, überhaupt von ihnen in Worten zu reden. Am meisten sind sie noch in der Aesthetik berücksichtigt worden, wo sie als "Anschaulichkeit", "unbewusste Vernunftmässigkeit", "sinnliche Verständlichkeit" und

The overwhelming superiority of propositional knowledge ("the kind of knowledge that has evolved to the level of linguistic articulation"[1]) consists first of all in that it transforms experience and knowledge into a social commodity that can be accumulated, tested and transmitted. Secondly, the possibility of organized cooperative action of human beings and thus the greater part of the species' specific strength is conditioned by language.

> In neither respect sensory knowledge can compete with propositional knowledge; yet this does not necessarily imply a lower degree of clarity or a different nature of sensory knowledge.[2]

5. The synthetic functions of subconscious mental operations according to 19th and 20th century theoretical developments. The problem of realism.

The doctrine of unconscious mental processes and their synthetic functions in perceptual integration was variously held by some other 19th century scholars. Thus, Mill invoked a *Law of Obliviscence* while arguing against W. Hamilton's thesis that "if we knew the parts before the whole, we must continue to know the parts better than the whole."[3] Mill states:

> It is one of the principal Laws of Obliviscence, that when a number of ideas suggest one another by association with such certainty and rapidity as to coalesce together in a group, all those members of the group which remain long without being specially attended to, have a tendency to drop out of conscious-

---

ähnlichen halbdunkelnen Bezeichnungen eine grosse Rolle spielen. Es steht ihnen das sehr falsche Vorurtheil entgegen, dass sie unklar, unbestimmt, nur halbbewusst vor sich gehen, dass sie als eine Art rein mechanischer Operationen dem bewussten und durch die Sprache ausdrückbaren Denken untergeordnet sind. Ich glaube nicht, dass in der Art der Thätigkeit selbst ein Unterschied zwischen den ersteren und den letzteren nachgewiesen werden kann." (VR I 361).

1 "[D]es bis zur Anwendung der Sprache gereiften Erkennens" (VR I 263; note the evolutionary perspective implicit in Helmholtz's phraseology).

2 "In beiden Beziehungen kann das Kennen nicht mit dem Wissen rivalisieren; doch folgt daraus nicht nothwendig eine geringere Klarheit oder eine andere Natur des Kennens." (VR I 362).

3 J.S. Mill, *An Examination of Sir William Hamilton's Philosophy,* (London, 1889^6), p. 323.

> ness. Our consciousness of them becomes more and more faint and evanescent, until no effort of attention can recall it into distinctness, or at last recall it at all.[1]

Indeed, James Mill had already suggested this idea.

> It not infrequently happens in our associated feelings, that the antecedent is of no importance farther than it introduces the consequent. In these cases, the consequent absorbs all the attention, and the antecedent is instantly forgotten.[2]

In this context it is interesting to note the phenomena of *backward masking* or *metacontrast* which have been extensively studied by Baxt, Pieron, Werner, Kolers and Rosner, Sperling, Lindsley and Emmons. Under certain specifiable conditions antecedent sensory stimulations, if presented briefly but long enough for correct identification under standard conditions, can be obliterated altogether from our consciousness if followed in a rapid sequence by a sufficiently intense stimulus which occurs within about 50 milliseconds of the former. Kolers has even suggested that such subliminal perceptions, though undetected, may influence behavior.[3]

Helmholtz very often admits, nay stresses, the fact frequently put forward by his opponents that the perceptual results which he attributes to mental and largely associative activities strike our consciousness with such apparent immediacy and urgency as though indeed produced by mere sensation. We find the same difficulty in Mill's account. For whatever the precise nature of these activities, one would expect them to consume time. This objection to the theory of unconscious inferences, however, seems to be refuted, if not—to use Lakatosian rhetoric—'turned into a decisive victory' by recent research which tends to show that complex percepts indeed do develop over time.[4]

Helmholtz obviously would have been quite pleased with the results of very recent investigations into binocular perception, all of which tend to show that some critical faculty is unconsciously operative in the visual system with a

---

1 Ibid.

2 J. Mill, *Analysis of the Phenomena of the Human Mind,* 2 vols. (London, 1829), I, p. 76.

3 P.A. Kolers, 'Subliminal stimulation in problem-solving', *Amer. J. Psychol.* 70 (1957), pp. 437-41; for a review of the literature, cf. D.H. Raab, 'Backward Masking', *Psychol. Bulletin* 60, 2 (1963), pp. 118-29.

4 Cf. G. Smith, 'Visual Perception: An Event over Time', *Psychol. Review* 64, 5 (1957), pp. 306-13.

remarkable capacity for decision making.[1] The resourcefulness of binocular perception is indeed astonishing. Bela Julesz of Bell Laboratories showed[2] that binocular depth perception may even occur in the absence of recognizable monocular cues. Extending his random-dot-stereogram method (in which a random-number generator delivers the coordinates for the points of light to be displayed in pairs on the oscilloscopes for the left and right eye respectively, while a computer adds depth information to the stream of random points), John Ross and John Hogben conducted studies which reveal that the perceptual system extracts depth and motion information from the visual input *even before we are conscious of what we see!*

> We must conclude that binocular perception has access to records of visual input that are independent of what we see. This represents a radical break with the commonsense view that what we see constitutes all the sense data from which higher perceptual processes develop a conception of the scene before our eyes. It appears there are records of visual input that can be consulted before anything at all is seen in order to determine the proper framework for perception... A number of puzzling visual effects...can be brought together by the assumption that perception must wait on the analysis of independent visual records before we are able to perceive.[3]

Another nineteenth century theoretician of the unconscious was Wundt, the champion of the empiricist-associationist tradition among experimental psychologists. He found that the perceptual integration of sensation elements generates *emergent* compounds whose attributes cannot be understood as the mere additive results of the attributes of these elements.

> Thus in the psychic resultants [e.g., spatial and temporal ideas] a principle expresses itself which in view of its effects we might also call a *Principle of Creative Synthesis*. Although this principle has been recognized for a long time within the context of the higher creative functions of the mind, yet by and

---

1 It has even been suggested that purely æsthetic grounds may sway the decision making process. Thus A.L. Austin reported to Darwin in 1877 about a curious discovery he had made. Having placed two photos of two different persons' faces in a stereoscope—the portraits being about the same sizes, and looking about the same direction—he found that the faces blended into one in a most remarkable manner, "producing in the case of some ladies' portraits, in every instance, a *decided improvement* in beauty" (quoted in John Ross, 'The Resources of Binocular Perception', *Sci. Am.* 234, 3 (March, 1976), p. 81.

2 Cf. Bela Julesz, 'Texture and Visual Perception', *Sci. Am.* (Febr., 1965).

3 John Ross (1976), p. 85.

large its relevance for the totality of the remaining mental processes has not been sufficiently...appreciated.[1]

The fact that 'the whole is more than the sum of its parts' was of course a theoretical problem for elementist psychologists like Wundt whose strategy consisted in analyzing all contents of actual experience into original elements of sensation. Thus arose a scientific system (later to be called *structuralism*) whose program of research aimed at describing the integrative processes by which mental compounds are gradually built up and at formulating the principles governing their operation.

By contrast, the Gestalt approach denied both the necessity and the appropriateness of analysis. Starting from unanalyzed phenomena as its basic data this strategy bypassed the problem of synthesis altogether. Instead, it simply assumed the existence of organizing *cortical* mechanisms responsible for the formation of structured wholes or *Gestalten*. Visual phenomena could be explained by postulating physiological restraints, such as preference for and distortion towards figures of 'simple' and 'closed' form. Thus Köhler's theory of *cortical isomorphism* assumed that visual forms were represented in the brain by similarly shaped electrical brain fields. These 'brain traces' were supposed to tend to form simple and closed shapes because of their physical properties; much as bubbles tend to become spheres, as this form has minimum potential energy.[2]

Thus, the dictum that the whole is more than the sum of its parts posed no special problem for the Gestaltists as it did for Wundt and the structuralists. But the Gestalt strategy, unlike the cognitive and activist strategy initiated by Helmholtz, imposed insoluble problems upon the theory of (perceptual) knowledge. For as it invoked physiological principles exerting *general* restraints, it implied that "visual 'organizations' and distortions are due to physical restraints and forces which will not in general be relevant to the logical problems the brain must solve to infer objects from sensory patterns and stored data".[3] It is this problem—the problem of rescuing epistemological realism despite a full-fledged information-theoretical account of perception implying the operation of assimilating functions mediating between

---

1 "In den psychischen Resultanten kommt auf diese Weise ein Prinzip zur Geltung, dass wir im Hinblick auf die entstehenden Wirkungen auch als ein *Prinzip schöpferischer Synthese* bezeichnen können. Für die höheren geistigen Schöpfungen längst anerkannt, ist es zumeist für die Gesamtheit der übrigen psychischen Vorgänge nicht zureichend gewürdigt...worden." [W. Wundt, *Grundriss der Psychologie*, (Leipzig, 1905), p. 399].

2 W. Köhler, 'Die physische Gestalten', (1920), W.D. Ellis (tr.), *A Source Book of Gestalt Psychology*, (London/New York, 1938), pp. 17-54.

3 R.L. Gregory (1974), p. XXXVIII.

perceiver and perceived—which preoccupied Helmholtz and which will concern us in the following concluding chapter of this book.

XI

THE EPISTEMOLOGICAL OUTCOME OF HELMHOLTZ'S NATURALISM. HYPOTHETICAL REALISM.

1. Helmholtz's novel theory of causality in its relation to Kant, Reid, and traditional empiricism.

Helmholtz's original elaborations of Kantian themes in conjunction with his naturalistic attempts to bridge the gap between professional philosophy and science and to combine their respective intellectual resources has made him the victim of serious misunderstandings, especially on the part of his disquieted philosophical critics. The scientist-turned-philosopher had become an uncommon and suspect figure in his days. And the unfamiliar chords Helmholtz struck, with fundamental tones vaguely reminiscent of Kant but with harsh empiricist overtones, bewildered the philosophers' *plenum* even more, like a blue grass fiddler contending to direct the Berlin Symphony Orchestra. Hence views were ascribed to him which he had never held and the irony of many of his philosophical exchanges was that he was forced to defend himself against the very conceptions he himself had most consistently criticized. Thus he had to reiterate against Land that throughout his life he had combated the assumption

> that empirical knowledge is acquired by simple importation
> or by counterfeit, and not by peculiar operations of the mind,
> solicited by various impulses from an unknown reality.[1]

This is almost pitiful. For the assertion of the indispensability of mental co-operation and of lower psychic activities in the development of percepts and concepts had indeed always been at the heart of Helmholtz's epistemology. And again, it had been precisely this insight which, on the scientific side, Helmholtz had been most concerned to defend and which had triggered off the hotly debated controversy among nineteenth century physiologists concerning the relative contribution of direct intuition vs. learning and experience to the conceptual process discussed above.

---

1 'The Origin and Meaning of geometrical Axioms-II', *Mind*, 3 (1978), p. 222.

Yet the source of these persistent misunderstandings can easily be detected. Helmholtz, for all his genuine recognition of the Kantian emphasis on a priori forms as the ultimate foundation of all perceptual knowledge, nevertheless made sustained and meticulous efforts to narrow down their actual range of influence. Even then he redefined the proper function of the remainder of these formal structures in terms that rendered them intelligible as well as scientifically scrutinizable. Thus, in spite of the undeniable fact that there are certain formal similarities between Helmholtz and Kant, the differences between them are definitely more conspicuous. Hamm is therefore right in concluding:

> Helmholtz seeks the foundation of knowledge...in natural science. Kant, by contrast, seeks the foundation of science in the apriori of knowledge.[1]

It might seem, however, that the category of causality constitutes the sole exception to Hamm's general conclusion. It is indeed the only category Helmholtz retains of all the apriori forms of pure reason listed by Kant. Its aprioricity is beyond doubt, Helmholtz consistently asserts, at least in his official writings. It contains the sole fundamental presupposition of any theory of knowledge. Unless a distinction is made between what is and what is not given to our consciousness and unless the sensations are interpreted as the *effects* wrought upon our senses by an independent reality, neither thought nor intentional action can occur at all and epistemological appraisals regarding the trustworthiness of our senses, of our perceptions and ideas, are neither possible nor called for. Prior to any experience an awareness must arise that the sensations do not constitute an autonomous imagery of the mind but that they are mediating links in a causal chain which connects the world of the mind with an independent world order.

Sensations by themselves do not constitute an experience. To perceive an object is radically different from merely having certain sensations. Galilei, if he had been hypnotized or drugged when he turned his telescope towards Jupiter, might never have *discovered* its satellites in spite of his having had all the luminous sensations needed for their discovery. Eyeballs and optic nerves are blind. Visual sensations by themselves, unless interpreted as signs of an external world order, are rather like "the vacant vista of aimless staring through a railway window".[2]

---

1 "Helmholtz begründet die Erkenntnis...naturwissenschaftlich. Kant dagegen begründet die Naturwissenschaft durch das Apriori der Erkenntnis." [Josef Hamm, 'Das philosophische Weltbild von Helmholtz', (Inaugural diss., Bielefeld, 1937), p. XXXIV].

2 N.R. Hanson, *Patterns of Discovery*, (Cambridge, 1969), p. 26.

In Fichtean terms, the Ego must first posit a non-Ego reality before the eye can see and the mind can exercise its intellectual functions.[1] This primordial 'quantum jump' from the level of phenomenal sensations to the level of unknown external agencies must have been afforded by the mind's original a priori structure. The postulate of causality expresses an unconscious mental operation entering into each and every experience and hence not derivable from the latter. Without it no experience would be possible. The world thus known to us can be interpreted only as a causal structure and we can tell about it only by its phenomenal effects. Its 'real' nature will for ever remain shrouded in uncertainty. The dogmas of materialism and idealism as well as any intermediate philosophical position can be tolerated only as metaphysical hypotheses. If taken in this latter sense and not as alleged necessities of thought they are legitimate and possibly even useful from a scientific point of view.[2]

When the *same object* appears now at this place, now at another, we must infer the operation of real causes generating the conception of the specific locality of a given object at a given time. Helmholtz calls them *topogenic* factors about whose nature we are entirely ignorant.[3] Again, since the *same place* at different times appears to our consciousness as occupied by different objects with different qualities there must be other causes operative in the real world which Helmholtz calls *hylogenic* factors whose nature will likewise remain unknown to us.[4] Thus all we can hope to learn about the world is its lawlike order as reflected by the regular temporal relations among the phenomenal effects it generates in our consciousness.

> We have in our language a very felicitous name for that which remains the same behind the flux of appearances and acts upon us, namely: "the actual". This expresses no more than the notion of action, without any supplementary reference to an enduring substance such as is implied by the concept of the real, i.e., of the material object.[5]

Whether we dream or not, whether the reality which gives rise to the phenomena consists of material objects or of immaterial entities or psychic states

1 EW 140.
2 EW 138.
3 EW 159-60; VR 402.
4 EW 160; VR 403.
5 "Wir haben in unserer Sprache eine sehr glückliche Bezeichnung für dieses, was hinter dem Wechsel der Erscheinungen stehend auf uns einwirkt, nämlich: "das Wirkliche". Hierin ist nur das Wirken ausgesagt; es fehlt die Nebenbeziehung auf das Bestehen als Substanz, welche den Begriff des Reellen, d.h. des Sachlichen, einschliesst." (SE 132).

and motives, the defining criterion of what we designate as *real* consists in no more than its lawlike causal effectiveness.

> The lawlike is...the essential presupposition of the character of the real.[1]

But if this is essentially what we mean when we speak of the 'real' world, then the noumenal reality can be known after all! For although the qualities of sensation are mere empty symbols without any qualitative similarity to the properties of the external world, they are nevertheless signs of *something*. The relation between the symbols and what they represent is only confined to this: that the same object operative under the same conditions will produce the same phenomenal sign and thus different signs must correspond to different causes. This rock-bottom relation—which is not a relation of qualitative resemblance, not even in part, but rather a formal relation of purely symbolic representation— makes the system of phenomenal signs into an extremely powerful cognitive tool. For, after all, this relation *does* copy external reality, if only its lawlike structural aspects, its regularities and the sequential order of its events.

> Thus, even though our sensations are *signs*, whose particular character entirely depends upon our own organization, yet they are not to be dismissed as a mere semblance. On the contrary, they are precisely signs of *something*, be it a substance or an event, and—most importantly—they can represent to us the *law* of this event.[2]

Now Helmholtz seems to argue that since the concept of external reality essentially involves no more than the totality of mind-independent events producing the phenomena which we are conscious of, and since regular relations among the latter accurately reflect regular relations among the former, we do have access to Kant's noumenal reality, at least in this its formal lawlike character. This 'minimal' or, as we shall see, hypothetical realism he maintains although he concurs with Kant that the attempt to conceive of reality *per se* in positive terms, untinged by our conceptual forms, is self-contradictory.

---

1 "Das Gesetzmässige ist...die wesentliche Voraussetzung für den Charakter des Wirklichen." (Ibid).

2 "Wenn also unsere Sinnesempfindungen *Zeichen* sind, deren besondere Art ganz von unsere Organisation abhängt, so sind sie doch nicht als leerer Schein zu verwerfen, sondern sie sind eben Zeichen von *Etwas*, sei es etwas Bestehendem oder Geschehendem, und was das Wichtigste ist, das *Gesetz* dieses Geschehens können sie uns abbilden." (SE 116).

> What we can attain, however, is an acquaintance with the lawlike order in the realm of the real, albeit in no other way than as represented in the sign-system of our sense impressions.[1]

Helmholtz's reasoning, as it stands, is of course not unobjectionable. Moreover, a correct understanding and emendation of it would, I think, involve considerable deviations from Kant's original doctrine of the a priori. Let us notice first of all that Helmholtz's realism is not the kind of transcendental realism envisaged by Kant but rather a kind of hypothetical scientific realism. On the one hand the external world can surely be known. On the other hand, any such knowledge is bound to ultimately rest on unverifiable hypotheses. Moreover, it is science alone which provides this reliable though hypothetical knowledge of the real world. To be sure, it is the peculiar activity of the mind to form general conceptions, that is, to search for causes.[2] But the intellect cannot simply impose at will any arbitrary causal structure upon the world.

> Thus the laws of nature confront us as an alien power, that cannot be disposed of at will.[3]

Elsewhere Helmholtz states:

> Laws of nature are...generic notions for the changes in nature.[4]

Thus whatever laws of nature we may venture to formulate are co-dependent on the nature of the mind as well as on the lawlike regularities among those phenomena which we recognize as not produced by conscious volitions and therefore as being independent of the formal properties of our mental and perceptual faculties. Such phenomena are to be categorized as natural phenomena. They are the objects of investigation for science. An early note published by Koenigsberger and according to him written in 1847 reads:

> The object of science is that aspect of our representations which we regard as not being generated by the autonomous activity of our faculty of representation, i.e., that which we perceive as actual [*scil.* as acting upon us].[5]

---

1 "Was wir aber erreichen können, ist die Kenntnis der gesetzlichen Ordnung im Reichen des Wirklichen, diese freilich nur dargestellt in dem Zeichensystem unserer Sinneseindrücke." (SE 132).

2 PO III 35.

3 "So treten uns die Naturgesetze gegenüber als eine fremde Macht, nicht willkürlich zu wählen." (VR I 375).

4 PO III 34.

5 "Naturwissenschaft hat zum Objekte denjenigen Inhalt unserer Vorstellungen, welcher von uns als nicht durch die Selbsttätigkeit unseres Vorstellungsvermögen

All those phenomena, therefore, which are not direct products of thought or imagination but present themselves to the mind rather as simply given, as *objecta*, are by definition 'real' phenomena in the sense of being generated by an objective reality whose precise qualitative nature we have no means of knowing. Knowledge of this reality cannot and does not consist in knowing its metaphysical nature *per se*. The use of abstract concepts without factual import merely obfuscates the facts. They remain meaningless unless they can be employed in drawing inferences regarding "new observable regular relations among the phenomena".[1] All we can significantly mean, then, by a complete knowledge of reality is a comprehensive understanding of its mechanisms, of how it works and how it produces regular changes. Surely, our mode of comprehension is determined by the mind's a priori category of causality, and the formulation of laws primarily expresses our method of thinking. Thus "[the law of sufficient reason] is not a law of nature".[2] On the other hand, such natural laws find their objective validation in the observed regular changes among those phenomena whose occurrence is felt to be independent of our mental activity. It is for this reason that we must ascribe such changes to natural causes, thereby objectifying the laws in question.

> Nature's lawlikeness is conceived of as causal connection as soon as we recognize its independence from our will.[3]

Of course, 'the laws of nature' we happen to formulate are bound to remain hypothetical to some degree due to our incomplete knowledge and the inductive character of the inferences involved in our observations of reality from the very beginning.[4] Yet,

> if it is found that the natural phenomena are to be subsumed under a definite causal connection, this is certainly an objectively valid fact, and corresponds to special objective relations between natural phenomena, which we express in our thinking as being their causal connection, simply because we do not know how else to express it.[5]

But how does this exposition square with the alleged aprioricity of the category of causality? It certainly removes the underpinnings for the common idealist

---

erzeugt angeschaut wird, d.h., also das als wirklich wahrgenommene." [L. Koenigsberger (1902-3), II, p. 126. Cf. VR II 242].

1 SE 133.

2 PO III 34.

3 "Die Gesetzlichkeit der Natur wird als causaler Zusammenhang aufgefasst, sobald wir die Unabhängigkeit derselben von unserem Willen anerkennen." (VR I 377).

4 SE 133.

5 PO III 34.

interpretation of Kant's critical philosophy which inspired Nietzsche's perspectivist theory of knowledge. But one may well ask whether it is still compatible with the Kantian doctrine of causality. Frankly, I do not see how it could be. For if the Law of Causation is a constitutive principle of human thought—as it is with Kant—then, indeed, it would seem to me, there can be no escape from the compelling argument that whatever world might be supposed to exist behind the appearances and our conceptions thereof, must for ever remain entirely incognizable. Indeed, Kant's transcendental adherence to the 'real' world in however qualified a sense may very well merely satisfy, on inconsistent grounds, a persistent and almost religious craving for recognizing a world beyond any human power to know. Yet, this kind of 'realism' cannot impress us. For to uphold the existence of a world which is simultaneously shown to be literally inconceivable is clearly incoherent. Nor is denying its existence, as the Neo-Kantian idealists did, any more intelligible. As Nietzsche observed,rather bluntly as usual:

> The *Ding an sich* is worth a Homeric laugh, since it *seems* to be so much, to be *everything*, when really it is empty—empty of meaning.[1]

At any rate, as long as the aprioricity of mental categories is taken as *constitutive* of the mind's operations, there is no room left for any correspondence theory of truth. It is no surprise, therefore, to find a definite development in Helmholtz's theory of causality towards a recognition of the Law of Causation as a merely *regulative* principle of the mind in need of, and indeed finding, continual corroboration in its pervasive and successful applicability. In *The Facts in Perception*, which contains the most comprehensive and mature exposition of Helmholtz's views, this gradual transition culminates in the assertion that the Law of Causation amounts to no more than a bold heuristic hypothesis concerning the complete intelligibility of the world whose validity, though not derivable from experience, is nevertheless tested by its factual success and not, as with Kant, a mere logical consequence of the necessary laws of human thought.[2] The assumption of the regularity of all natural phenomena, which we must make in order that we may understand corporeal reality, is still no more than a primordial hypothesis. As such it expresses our particular mode of understanding the world. It only differs from any other hypothesis insofar as we cannot intelligibly formulate any rival hypothesis while continuing to understand the world, and act in it, on this novel basis. In Lakatosian terms, the causal law is made irrefutable by *fiat* since we feed

1 Quoted in Arthur Danto, *Nietzsche as Philosopher*, (New York, 1965), p. 83; from *Menschliches, Allzumenschliches*, (1878), p. 20.

2 TW, in SE 133.

it into the negative heuristic of each and every scientific research program while their respective 'protective belts' of additional hypotheses must bear the brunt of apparent anomalies. It is *not* irrefutable, however, in the sense that we cannot conceive of an anomalous world to which it could be applied with only limited success, or even with no success at all.[1] In fact, the neo-vitalistic hypothesis to which Helmholtz's celebrated *The Conservation of Energy* had dealt the final blow, provides just one example (among many others in the history of ideas) where a large section of the real world is assumed to be exempt from the universal reign of strict causality.[2]

The increasingly hypothetical import in Helmholtz's changing interpretation of the idea of causality transpires clearly from a late note published by Koenigsberger:

> The causal law (the presupposition of the lawlikeness of nature) is only a hypothesis and not otherwise provable. Lawlikeness in the past can never prove lawlikeness in the future. The only proof of all hypotheses is always: test whether it holds, and you will find it (preferably by experiment, when applicable). In contrast to other hypotheses expressing particular laws of nature the causal law occupies an exceptional position only in the following sense: 1. It is the presupposition of the validity of all the others. 2. It alone gives us the possibility in principle to know something unobserved. 3. It is the necessary basis for intentional acts. 4. We are driven towards it by the natural mechanism of our representational network. Thus we are driven by the strongest motives to wish it true; it is the basis of all thinking and acting. Unless we have it, we cannot even begin to test it; we must anticipate its success, then the success will count as confirmation. We must be aware that we expected the success, then we will become aware of the law. Thinking means seeking the rule of law; judging means one has found it. Thus, without the causal law no thinking is possible. No thought without recognition of the causal law is therefore a tautology; one may ask whether our thinking is justified and whether it makes any sense; whether it does can only be proved by (inner or outer) acting.[3]

---

1 *Über die Erhaltung der Kraft*, SE 133.

2 VR I 385.

3 " Das Causalgesetz (die vorausgesetzte Gesetzmässigkeit der Natur) ist nur eine Hypothese und nicht anders erweisbar als solche. Keine bisherige Gesetzmässigkeit kann künftige Gesetzmässigkeit erweisen. Der einzige Beweis aller Hypothesen ist immer:

Commenting on this passage Erdmann concludes:

> ...the tendency to transform the Kantian apriori into an empiricist train of thought...is unmistakable. In this respect his [Helmholtz's] view resembles the philosophy of Hume (whom he has seen in the light of his time as no more than a sceptic), as well as the ideas of Stuart Mill rather than those of Kant.[1]

However, this characterization is inadequate. It leaves out the more interesting aspects of both the similarities and dissimilarities of Helmholtz's relationship to his philosophical environment. Helmholtz never abandoned the Kantian notion of the aprioricity of the idea of causality. Quite similarly, however, he continued to adhere, as we have seen, to the aprioricity of our spatial apperceptions, but in a sense that was radically different from Kant. What made him a lifelong disciple of Kant was his emphasis on the indispensability of *mediating* intellectual and psychic processes for the production of reliable knowledge of reality. However, contrary to Kant Helmholtz did not conceive of these mediating processes as necessarily restrictive fixed structures, but rather as goal-directed adaptive mechanisms. They are best characterized as *cognitive strategies* for tackling information-theoretical problems confronting the perceptual system in the course of its development. As such they are not fixed attributes of a fixed mind creating a world of their

---

prüfe, ob es so ist, und Du wirst es finden (am besten experimentell, wo es angeht). Den übrigen Hypothesen, welche besondere Naturgesetzen aussagen, gegenüber hat das Causalgesetz nur folgende Ausnahmestellung: 1. Es ist die Voraussetzung der Gültigkeit aller anderen. 2. Es giebt die einzige Möglichkeit für uns überhaupt, etwas nicht Beobachtetes zu wissen. 3. Es ist die nothwendige Grundlage für absichtliches Handeln. 4. Wir werden darauf hingetrieben durch die natürliche Mechanik unserer Vorstellungsverbindungen. Wir sind also durch die stärksten Triebfedern getrieben, es richtig zu wünschen; es ist die Grundlage alles Denkens und Handelns. Ehe wir es nicht haben, können wir es auch nicht prüfen; wir müssen den Erfolg voraus denken, dann ist der Erfolg eine Bestätigung. Wir müssen uns bewusst sein, dass wir den Erfolg voraus erwartet haben, dann werden wir des Gesetzes bewusst. Denken heisst die Gesetzmässigkeit suchen; urtheilen heisst sie gefunden haben. Ohne Causalgesetz also kein Denken. Kein Denken ohne Anerkennung des Causalgesetzes ist also eine Tautologie; es fragt sich, ob wir zum Denken berechtigt sind und ob das einen Sinn hat; dieser Sinn lässt sich nur durch die Handlung (innere oder äussere) erweisen." [Koenigsberger (1902), I, pp. 247 f].

1 " ...der Zug zur Umbildung des kantischen Apriori in einen empiristischen Gedankengang...ist unverkennbar. Seine [Helmholtz's] Meinung ist insoweit der Lehre Humes, den er im Lichte seiner Zeit lediglich als Skeptiker gesehen hat, sowie den Gedanken Stuart Mill's ähnlicher als denen Kants." [Benno Erdmann, 'Die Philosophischen Grundlagen von Helmholtz' Wahrnehmungstheorie', in *Abhandlungen der Preussischen Akademie der Wissenschaften*, Philosophisch-Historische Klasse I, (Berlin, 1921), p. 11 f].

own (the purely phenomenal and in principle intelligible world) in contradistinction to the 'real' (or 'noumenal') world which by definition can never be known or understood in positive terms. On the contrary, these strategies are unstable and flexible structures, hypothetical perspectives subject to feed-back control and thus providing the means for intelligent interaction with any possible environmental input.

In Kant the a priori forms have no more than a purely assimilative function. They cannot serve any other purpose. Without them no input could be experienced and no experience could be understood. Furthermore, the otherwise baffling harmony between mind and nature is likewise to be explained as a function of the inevitable assimilating operations of man's perceptual and intellectual faculties. By thus solving the riddle of the objective validity of primitive notions and propositions in terms of which all knowledge must ultimately be justified Kant provides the master-theory for the rationalist tradition in the theory of knowledge. But he pays a considerable price. For the alleged fit between mind and world can have no more than a merely intraphenomenal validity. And while 'realism' is upheld in name, reality reduces to the shadowy realm of the ineffable *Ding an sich*, a kind of philosophical black hole from which no intelligible signal will ever escape. Thus Reason, just because it is active and structured (like a block of marble with veins rather than like an absolutely clean slate) must irredeemably stand between itself and its ultimate objective: to obtain knowledge of reality as it is in itself. Heisenberg's indeterminacy relation for the micro-structure of the physical world is merely a particular case of the more radical indeterminacy Kant attributes to the entire 'real' world: the phenomena are relative not just to a particular observational situation, nor merely to a particular observer, not even to the knowing agent in question, but to the necessary features of human knowledge as such.

This Kantian doctrine, which was not unjustly taken to be the doctrine of the relativity of all human knowledge (notwithstanding the objective validity of its necessary formal foundations), profoundly scandalized the nineteenth century, the more so in philosophical circles which were generally impressed with other tenets of the Kantian system. As indicated above, this provoked and stimulated philosophical thought in very different directions. Thus Hegel's dynamic rationalism attempted to rescue absolute knowledge conceived as the state upon which the History of Reason through its own inner dynamics and continual antagonisms would ultimately converge. Helmholtz, by contrast, conceived of absolute knowledge (or knowledge of the real world) as the state which the conjecturing mind of the individual percipient gradually converges upon through a subconscious interpretive process of successive object hypotheses on the basis of purely structural relations between empty uninterpreted signs or symbols of experience. Thus he developed a learning theory for subconscious data which, as we have noted, represented a considerable advance over eighteenth century psychological theory. Secondly, while he was in agreement with Reid that what we perceive is indeed objects (and not ideas or sensations) he maintained, contrary to Reid (and to Kant), that the epistemic transition from sensation to perception was due not to fixed principles of our human constitution nor (contrary to Hume) to natural imagination but to pragmatically controlled hypotheses subject to change.

Thus, because on the level of *psychology* he regarded perception as a hermeneutic, creative, and even artistic enterprise drawing on largely subconscious intellectual resources, on the level of *epistemology* Helmholtz managed to maintain realism. At the same time his psychological theory enabled him to steer a delicate course between Reid's dogmatism and Kant's idealism, and to avoid both direct realism à la Reid and phenomenal realism à la Kant. As a result, Helmholtz maintained a hypothetical brand of realism. For contrary to his predecessors experience in Helmholtz is the joint achievement of assimilating as well as accommodating functions. Therefore no pre-established harmony needs to be presupposed or argued for. Rather, the mind gradually 'tunes in' to the structures of reality as it develops, by small leaps of analogy, cognitive strategies which are progressively more appropriate for dealing with increasingly complex environmental inputs and tasks. Consequently the resulting harmony holds between the *real* world and our conceptions thereof, and not merely (as in Kant) between reason on the one hand and a formally ordered and a priori conceptualized reality on the other. Nor is the harmony produced by the benevolence of God (as in Reid). On the contrary, it is the fruit of man's intelligent interaction with his environment. This real harmony can be attained only through a process of *learning*. This is the crucial concept in Helmholtz's 'genetic' epistemology. And it is this emphasis on genuine mental activity and development which distinguishes his theory from classical empiricism as well as from classical rationalism and Kantian apriorism. His so-called *empirical* theory is in fact a *cognitive-interactionist* theory of perception. And just as the notion of genuine mental activity would later provide Piaget with the epistemological link explaining the success of what would otherwise be an arbitrary play of the mind and its successive equilibria, so the notion of unconscious inferences, of subliminal but intelligent strategies of information-processing, enabled Helmholtz to maintain hypothetical realism. All that is missing is the concept of *structuralism*. To that extent

Helmholtz is indeed more akin to the empiricist tradition than to the Kantian one.[1]

2. Lack of an adequate psychology. Weaknesses of Helmholtz's theory.

As a consequence of his epistemological position Helmholtz, in his investigation into the physiological and psychological factors operative in our perceptual transactions with the world, naturally allowed little scope to innate fixed structures. This explains on one extreme of the epistemological spectrum his sustained and well-supported criticism of the intuitionist physiologists whose research program aimed at deriving all (perceptual) knowledge from immediate experience by assuming direct input-output links in a way that is in this respect paradoxically enough substantially similar to the sensationalist philosophers who endeavored, in their opposition to the rationalist views of their time, to establish all human understanding as derivable from a strictly empirical basis alone by means of an atomistic psychology.[2] Helmholtz is squarely opposed to this view:

> The interpretation of our sense-impressions is founded upon experiment and not upon mere observation of external events.[3]

On the other end of the epistemological spectrum, however, this same aversion from innate contrivances or fixed structures necessarily determining and restricting human knowledge explains Helmholtz's growing discomfort with the rationalistic aspects of the Kantian system, whose a priori forms he came to regard as too rigid and even, contrary to Kant's intentions, as too rich in content to allow for all possible experiences. The attack he leveled (which he himself regarded as a modest emendation rather than as a fundamental critique) proved successful at least in the case of the philosophy of mathematics. However, as was pointed out above, his general reservations concerning

1 This allows an interesting comparison with Quine's naturalistic epistemology which also lacks structuralist concepts (except on the relatively high level of scientific discourse), but which similarly employs the notion of a progressive construction of the world through "short leaps of analogy". Cf. W.V.O. Quine, 'The Nature of Natural Knowledge', in *Mind and Language*, S. Guttenplan (ed.), (Oxford, 1975), p. 78.

2 PO III 35.

3 "Die Deutung unserer Sinnesempfindungen beruht auf dem Experiment und nicht auf blosser Beobachtung aüsserer Geschehens." (VR I 355).

the restrictive role Kant had attributed to the mind's mediating functions could not so easily be turned into a philosophical victory in the case of the idea of causality. Helmholtz had dropped the other a priori forms of pure reason as unnecessary. But the assumption of the causal law as a necessary law of thought appeared indeed to be indispensable for an adequate explanation of perception. For we do not ordinarily perceive our sensations. Rather, what appears in most of our perceptions are qualities of external objects in time and space, their existence, size and location.

The problem of causality was more difficult for Helmholtz to crack than the riddle of how the general concept of space can be developed on the basis of primitive mental structures, or the question of how mature vision can attain the incredible precision achieved in its localization of particular objects. For one thing, the category of causality, unlike the concept of space, is not a perceptual form but a purely intellectual function. Thus, in dealing with this category Helmholtz the physiologist of perception found himself upon unfamiliar territory. With regard to the development of spatial apperceptions the epistemologist could heavily lean on his physiological knowledge and his mathematical skills. These enabled him to separate out the sensational input, or 'the given', in perception and thus to isolate the contributions made by lower psychic and intellectual activities.

> I believe we must regard the analysis of the concept of intuition into the elementary processes of thought as the most essential progress of the modern era. This analysis was still failing in Kant and its absence has been responsible for his [incorrect] conception of the axioms of geometry as transcendental propositions. Due to physiological investigations into sensory perception in particular we have been brought to the ultimate elementary processes of thought, processes inexpressible in words and thus doomed to remain incognizable and inaccessible to philosophical inquiry as long as the latter restricted itself to an analysis only of knowledge as expressed in language.[1]

---

1 "Als wesentlichsten Fortschritt der neueren Zeit glaube ich die Auflösung des Begriffs der Anschauung in die elementaren Vorgänge des Denkens betrachten zu müssen, die bei Kant noch fehlt, wodurch dann auch seine Auffassung der Axiome der Geometrie als transzendentale Sätze bedingt ist. Es sind hier namentlich die physiologischen Untersuchungen über die Sinneswahrnehmungen gewesen, welche uns an die letzten elementaren Vorgänge des Erkennens hingeführt haben, die noch nicht in Worte fassbar, der Philosophie unbekannt und unzugänglich bleiben mussten, so lange diese nur die in der Sprache ihren Ausdruck findenden Erkenntnisse untersuchte." (SE 134)

Physiological studies, however, could be of no avail to resolve the problem of the origins and the development of the modal concept of causality. If empirical research were to be relevant at all, it would have to be research strictly within the area of cognitive and developmental psychology. And although Helmholtz no doubt gave a great impetus to the development of these fields of study, his own contribution to it was confined to theoretical suggestions and groundwork not involving actual research.

Thus Helmholtz found himself faced with an awkward dilemma. If he accepted Kant's doctrine of causality as a constitutive principle of the mind, he could not consistently uphold his scientific realism which he had always been at pains to defend, unless he was also willing to accept some version of the doctrine of pre-established harmony between the real world and the laws of thought, a metaphysical doctrine, that is, of which he himself had always been highly critical. A Spencerian solution, which applied Darwin's theory of natural selection to the Kantian categories and according to which these categories are indeed a priori to the individual but nevertheless a posteriori to the species, never appealed to him in spite of his frequent recognition of Darwin's achievement and, even more surprisingly, in spite of the parallel he occasionally drew between the adaptive potential of the individual organism as championed in his own theory of perception and the adaptive features on the level of phylogenetic development as hypothesized in Darwin's theory.

> Darwin's theory contains an essentially novel, creative idea. It demonstrates how the morphological adaptedness of organisms can also be due to the blind operation of natural law without any ancillary intelligence.[1]
>
> ...Whereas Darwin's theory exclusively focuses on the gradual evolution of species as it occurs in the succession of generations, it is well known that the individual by itself also adapts to, or as we are wont to say, accustoms itself to a certain extent to its ecological conditions; that is to say, that even during the individual's lifetime a certain measure of higher development can be gained. And precisely in that area of organic life whose forms have reached the highest degree of adaptedness and reaped the greatest admiration, viz. in the area of sensory perception, the latest advances of physiology

---

1 " Darwins Theorie enthält einen wesentlich neuen schöpferischen Gedanken. Sie zeigt, wie Zweckmässigkeit der Bildung in den Organismen auch ohne alle Einmischung von Intelligenz durch das blinde Walten eines Naturgesetzes entstehen kann." (VR I 388).

inform us that this individual adaptation plays a very prominent role.[1]

Yet, in spite of these insights Helmholtz never admitted a Darwinian compromise between the only two alternatives he was willing to consider as possible explanations of the epistemological riddle of the apparent harmony between the inner and the outer world: this harmony is either "a prepared product of organic creativity"[2] or else it is "an individually acquired adaptation,... a product of experience".[3]

The other horn of the dilemma, though more in conformity with Helmholtz's realistic convictions, was hardly more promising. For if he rejected the Kantian idea of the aprioricity of the category of causation in an attempt to show that our fundamental causal belief is somehow attuned to the real world, he would have to supply a well-supported descriptive account of how such a belief could have been developed on the basis of more primitive structures. This, however, would require elaborate psychological studies (as was recognized soon thereafter[4]) for which, however, he lacked both the research-skills and the facilities. Moreover, the very idea of 'nursery psychology' and its empirical study had not yet arisen but arrived only towards the end of the nineteenth century.[5] Psychology used to be faculty psychology, not functional psychology, and its method was introspection, not empirical analysis.[6]

This lack of a sufficiently corroborated psychological theory of the development of causal schemes made Helmholtz's systematic attempt at naturaliz-

---

1 "...Während Darwins Theorie sich ausschliesslich auf die durch die Reihe der geschlechtlichen Zeugungen eintretende allmähliche Umformung der Arten bezieht, ist bekannt, dass auch das einzelne Individuum sich den Bedingungen, unter denen es zu leben hat, bis zu einem gewissen Grade anpasst, oder, wie wir zu sagen pflegen, eingewöhnt; dass also auch noch während des einzelnen Lebens eines Individuums eine gewisse höhere Ausbildung der organischen Zweckmässigkeit seiner Bildungen den höchsten Grad erreicht und die meiste Bewunderung erlangt hat, nämlich im Gebiete der Sinneswahrnehmungen, lehren die neueren Fortschritte der Physiologie, dass diese individuelle Anpassung eine ganz hervorragende Rolle spielt." (VR I 390-1).

2 "ein vorbereitetes Product der organische Schöpfungskraft" (VR I 391).

3 "eine individuell erworbene Anpassung,... ein Product der Erfahrung" (VR I 394).

4 "The study of children is generally the only means of testing the truth of our mental analyses", Baldwin observed. [J.M. Baldwin, *Mental Development in the Child and the Race*, (London, 1900), p. 5].

5 Baldwin complains in 1894: "Now that this genetic conception has arrived, it is astonishing that it did not arrive sooner, and it is astonishing that the 'new' psychology has hitherto made so little use of it." (Ibid., p. 3).

6 VR I 111.

ing Kant's epistemology much less consistent with regard to the category of causality than with respect to his radical modification of the Kantian doctrine of perception where his epistemological insights could draw upon his physiological findings. Thus Helmholtz's theory of causality became a rather incoherent mixture of Kantian and empiricist elements. On the one hand, following Kant, the causal axiom must be valid prior to all inner and outer experience. In 1855 he writes:

> But how then did we ever manage for the first time to transfer from the world of our nerve impressions to the real world? Apparently only by means of an inference; we have to presuppose the existence of external objects as *causes* of our nerve excitations; for there cannot be any effect without a cause. How do we know that there cannot be any effect without a cause? Is that an empirical proposition? Indeed, it has been passed off as such. However, it is evident that we need this proposition even before we have obtained any knowledge whatsoever about the objects of the external world; we need it in order to arrive at all at the insight that there are objects in ambient space which can be related to each other as cause and effect. Can we then derive that proposition from the inner experience of our self-consciousness? No; because the self-conscious acts of our willing and thinking we regard precisely as *free*; that is, we deny that they are necessary effects of sufficient causes. Thus the investigation of sensory perception also leads us to the insight already found by Kant: that the proposition *'No effect without a cause'* is a law of our thought prior to all experience.[1]

---

1 "Auf welche Weise sind wir denn nun zuerst aus der Welt der Empfindungen unserer Nerven hinübergelangt in die Welt der Wirklichkeit? Offenbar nur durch einen Schluss; wir müssen die Gegenwart aüsserer Objekte als *Ursache* unserer Nervenerregung voraussetzen; denn es kann keine Wirkung ohne Ursache sein. Woher wissen wir, dass keine Wirkung ohne Ursache sein könne? Ist das ein Erfahrungssatz? Man hat ihn dafür ausgeben wollen; aber wie man sieht, brauchen wir diesen Satz, ehe wir noch irgend eine Kenntnis von der Dingen der Aussenwelt haben; wir brauchen ihn, um nur überhaupt zu der Erkenntnis zu kommen, dass es Objecte im Raume um uns gibt, zwischen denen ein Verhältniss von Ursache und Wirkung bestehen kann. Können wir ihn aus der inneren Erfahrung unseres Selbstbewusstseins hernehmen? Nein; denn die selbstbewussten Akte unseres Willens und Denkens betrachten wir gerade als *frei*; d.h. wir leugnen, dass sie nothwendige Wirkungen ausreichender Ursachen seien. Also führt uns die Untersuchung der Sinneswahrnehmungen auch noch zu der schon von Kant gefundenen Erkenntnis: dass der Satz: "Keine Wirkung ohne Ursache", ein vor aller

But even in this early lecture his apparent allegiance to Kant is clouded with obscurities. He has shown that mere physiology is insufficient to explain our perceptions: we also need psychology. However, introspective psychology would obscure rather than clarify the facts of perception.

> It is hard to determine the nature of the mental processes which transform the sensation of light into a perception of the external world. Unfortunately, psychology has no help to offer, since up till now psychology has used introspection as the only method for obtaining knowledge, whereas in this case we are concerned with mental operations about which introspection is utterly silent and whose existence is to be inferred, rather, from physiological investigations of the organs of sense.[1]

Helmholtz then goes on to develop his theory of unconscious inferences which specifies the nature of the mediating psychic processes involved in our perceptions. These inferences are carried out mechanically, "without awareness on our part and not subject to control by selfconscious intelligence".[2] Among other things, this theory serves to explain the phenomenon of persistent perceptual illusions. Is it also meant to explain the development of our belief in an external world as the cause of our sensations?

The above quote asserting the aprioricity of the causal axiom seems to preclude such a possibility. However, two pages before the passage quoted Helmholtz seems to present his theory of unconscious inferences precisely as a theory of induction by which we infer the existence of unobserved causes from their observed effects. Here his line of analysis seems altogether Humean.

> However, when the mind does not directly perceive the objects at their own place, it can only come to know them by way of inference. Because it is only through inferences that we can come to know what we do not directly perceive. I agree that we are not aware of drawing this inference. Rather, it is a

---

Erfahrung gegebenes Gesetz unseres Denkens sei." [VR I 115-6; again, the use of the correlative terms '*Wirkung*' ('*effect*') and '*Ursache*' ('*cause*') is infelicitous, of course].

1 "Die Natur der psychischen Prozesse zu bestimmen, welche die Lichtempfindung in eine Wahrnehmung der Aussenwelt verwandeln, ist eine schwere Aufgabe. Leider finden wir bei den Psychologen keine Hülfe, weil für die Psychologie die Selbstbeobachtung bisher der einzige Weg des Erkennens gewesen ist, wir es aber hier mit geistigen Thätigkeiten zu thun haben, von denen uns die Selbstbeobachtung gar keine Kunde giebt, deren Dasein wir vielmehr erst aus der physiologischen Untersuchung der Sinneswerkzeuge schliessen können." (VR I 111).

2 "[O]hne Selbstbewusstsein und nicht unterworfen der Controlle der selbstbewussten Intelligenz" (VR I 110).

> mechanically conditioned inference, that has emerged in the series of involuntary associations of ideas as these tend to arise whenever two representations have very frequently occurred in connection with each other. As a result they always evoke one another with a certain degree of natural necessity.[1]

Helmholtz tries to reconcile these conflicting accounts by distinguishing between the process connecting our sense impressions with the conception of an object of definite form and at a definite location—a process which he regards as the result of habitual association; and on the other hand the process by which we come to conceive of such external objects in the first place—a process which cannot be thus interpreted, and which therefore requires the causal axiom as its logical point of departure.

This line of thought, of course, is far from convincing. The philosophical foundations, then, of Helmholtz's theory of causality are certainly in need of further refinement and elaboration. But the overall naturalistic direction of research and the epistemological position of hypothetical realism made possible by it are altogether new and promising. The notion that realism could be saved, paradoxically, by regarding all perceptions essentially as *brain fictions* based on past experience, selected by present sensory data and guided by pragmatically controlled cognitive strategies would bear considerable fruit in twentieth century psychology and philosophy of perception, even though the philosophical and theoretical implications of the core ideas of Helmholtz's research program are being appreciated only in very recent times, almost a century since they were first fed into the 'meme pool' of Western thought.

---

1 "Wenn aber das Bewusstssein nicht unmittelbar am Orte der Körper selbst diese wahrnimmt, so kann es nur durch einen Schluss zu ihrer Kenntniss kommen. Denn nur durch Schlüsse können wir überhaupt das erkennen, was wir nicht unmittelbar wahrnehmen. Dass es nicht ein mit Selbstbewusstsein vollzogener Schluss sei, darüber sind wir einig. Vielmehr hat er mehr den Charakter eines mechanisch eingeübten, der in die Reihe der unwillkürlichen Ideenverbindungen eingetreten ist, wie solche zu entstehen pflegen, wenn zwei Vorstellungen sehr haüfig mit einander verbunden vorgekommen sind. Dann ruft jedesmal die eine mit einer gewissen Naturnothwendigkeit die andere hervor." (VR I 112).

BIBLIOGRAPHY

ABRAMS, M., *The Mirror and the Lamp, Romantic Theory and the Critical Tradition*, (New York, 1958).

ALHAZEN, *Optics*, MS Fatih 3212, tr. and quoted from the Arabic text by A.I. Sabra, 'Sensation and Inference in Alhazen's Theory of Visual Perception', in *Studies in Perception*, P.K. Machamer & R.G. Turnbull (eds.), (Columbus, 1978); also translated into a Latin version, known in the Middle Ages as *Perspectiva* or *De aspectibus*, and later published in a volume bearing the collective title *Opticae thesaurus, Alhazeni Arabis libri septem...*, [Basel (Risner), 1572; repr., New York, 1972].

ALKINDI,

- *Alkindus de radiis stellarum*, Digby 91, 16th century, fols. 66-80; alternative titles: *De radiis stellicis*, or *De radiis stellatis*.
- *De aspectibus*, in 'Alkindi, Tideus and Pseudo-Euklid. Drei optische Werke', A.A. Björnbo & S. Vogl (eds.), *Abhandlungen zur Geschichte der mathematischen Wissenschaften*, vol. 26, 3, (Leipzig/Berlin, 1912).

ANDERSON, B.F., *Cognitive Psychology*, (New York, 1975).

AQUINAS, Thomas, *Summa theologiae*. Latin text and English translation. Introduction, notes, appendices and glossaries, I-LX, (London, Blackfriars, 1964-1976).

ARISTOTLE, *The Student's Oxford Aristotle*, W.D. Ross (ed.), (New York, 1942).

AVICENNA, *Liber canonis de medicinibus cordialibus*, (Venice, 1555).

AYER, A.J., *The Foundations of Empirical Knowledge*, (London, 1940).

BACON, F., *Advancement of Learning*, with a special introduction by J.F. Creighton, (New York, 1944).

BACON, R., *The Opus Majus of Roger Bacon*, John H. Bridges (ed.), 2 vols. (Oxford, 1897; Frankfurt/Main, unveränderter Nachdruck, 1964).

BALDWIN, J.M., *Mental Development in the Child and the Race*, (London, 1900).

BAUER, H., 'Die Psychologie Alhazens auf Grund von Alhazens Optik dargestellt', *Beiträge zur Geschichte der Philosophie des Mittelalters*, vol. X, no. 5 (Münster, 1911).

BEARE, J.I., *Greek Theories of Elementary Cognition from Alcmaeon to Aristotle*, (Oxford, 1906).

BECHLER, Z., 'Newton's 1672 Optical Controversies: A Study in the Grammar of Scientific Dissent', in *The Interaction between Science and Philosophy*, Y. Elkana (ed.), (Atlantic Highlands, 1974), pp. 115-42.

BELL, C., *Idea of a New Anatomy of the Brain*, (London, 1811).
BERKELEY, G.,
- *The Theory of Vision Vindicated*, (London, 1733), A.A. Luce & T.E. Jessop (eds.), *The Works of George Berkeley, Bishop of Cloyne*, (London, 1948-57).
- *A New Theory of Vision, and Other Writings*, with an introduction by A.D. Lindsay, (Everyman's Library 483, London, 1957).

BERNARD, F., 'Thèse sur l'absorption de la lumière par les milieux non crystallisés', *Ann. de chim.* 3 (1852), XXXV, pp. 385-438.
BLUMENBERG, H., 'Pseudoplatonismen in der Naturwissenschaft der frühen Neuzeit', *Akademie der Wissenschaften und der Literatur, Abh. der Geistes- und Sozialwissenschaftlichen Klasse*, (Mainz, 1971), nr. 1, pp. 3-34.
BÖHMER, H., *Geschichte der Entwicklung der naturwissenschaftlichen Weltanschauung in Deutschland*, (Gotha, 1872).
BORING, E.G., *Sensation and Perception in the History of Experimental Psychology*, (New York, 1942).
BRÜCKE, E., 'Über die stereoskopischen Erscheinungen', in *J. Müllers Arch. für Anat. und Physiol.* (1841).
BÜCHNER, L.,
- *Sechs Vorlesungen über die Darwin'sche Theorie,...* (Leipzig, 1872), (1st ed. 1868).
- *Aus Natur und Wissenschaft*, (Leipzig, 1874).

BURIDAN, J., *Questiones et decisiones physicales insignium virorum*, G. Lockert (imp.), (Paris, 1518).
BUTTS, R.E., 'Whewell's Logic of Induction', in *Foundations of Scientific Method: The Nineteenth Century*, R.N. Giere & R.S. Westfall (eds.), (Indiana, 1973), pp. 53-85.
CALLAHAN, J.J., 'The Curvature of Space in a Finite Universe', *Scientific American*, 23, 2, (Aug. 1976), pp. 90-100.
CAMPBELL, D.T., 'Methodological Suggestions from a Comparative Psychology of Knowledge Processes', *Inquiry* 2 (1959).
CARNAP, R., *Der logische Aufbau der Welt*, (Berlin, 1928).
CASSIRER, R., *Das Erkenntnisproblem in der Philosophie und Wissenschaft der neueren Zeit*, (Stuttgart, 1957).
CATON, H., *The Origin of Subjectivity. An Essay on Descartes*, (New Haven, 1973).
CHAITIN, G.J., 'Randomness and Mathematical Proof', *Scientific American*, (May 1975) 232, 5, pp. 47 ff.
CLARKE, D.M., *Descartes' Philosophy of Science*, (University Park, 1982).
CLARKE, E., 'The Doctrine of the Hollow Nerve in the Seventeenth and Eighteenth Centuries', in *Medicine, Science, and Culture: Historical Essays in*

Honor of Owsei Temkin, L.G. Stevenson & R.P. Multhauf (eds.), (Baltimore, 1968), pp. 123-41.

COHEN, M.R. & DRABKIN, I.E. (eds.), *A Source Book in Greek Science*, (Cambridge, 1966).

CONRAT, F., *Hermann von Helmholtz' psychologische Anschauungen*, (Halle a.d. S., 1904).

CRANEFIELD, P., 'The Organic Physics of 1847 and the Biophysics of Today', *Journal Hist. Medicine* 12 (1957), pp. 407-23.

CROMBIE, A.C.,

- *Robert Grosseteste and the Origins of Experimental Science 1100-1700*, (Oxford 1953, 1962).
- 'The Relevance of the Middle Ages to the Scientific Movement', in *Perspectives in Medieval History*, K.F. Drew & F.S. Lear (eds.), (Chicago, 1964).
- 'The Mechanistic Hypothesis and the Scientific Study of Vision: Some Optical Ideas as a Background to the Invention of the Microscope', in *Historical Aspects of Microscopy, Papers read at a one-day conference held by the Royal Microscopical Society at Oxford, 18 March, 1966*, S. Bradbury & G.L.E. Turner (eds.), (Cambridge, 1967).

CULOTTA, C.A., 'German Biophysics, Objective Knowledge and Romanticism', in *Historical Studies in the Physical Sciences*, Russell McCormmach (ed.), 4 (1974), pp. 3-38.

DAMIANOS, *Schrift über Optik*, R. Schoene (ed.), (Berlin, 1897).

DANTO, A., *Nietzsche as Philosopher*, (New York, 1965).

DAVIDSON, A., 'Mental Events', in *Experience and Theory*, L. Foster & J.W. Swanson (eds.), (Mass. University Press, 1970).

DEMBER, W.N., *Visual Perception: The Nineteenth Century*, (New York, 1964).

DESCARTES, R.,

- *Oeuvres de Descartes*, C. Adam & P. Tannery (eds.), 12 vols., (Paris, 1897-1910).
- *The Philosophical Works of Descartes*, E.S. Haldane & G.R.T. Ross (trs.), 2 vols., (Cambridge, 1911-1912); corrected edition 1931; paperback ed. (New York, 1973).
- *Descartes: Selections*, R.M. Eaton (ed.), (New York, 1927).
- *Discourse on Method, Optics, Geometry, and Metereology*, P.J. Olscamp (tr.), (Indianapolis, 1965).
- *Descartes, Philosophical Writings*, E. Anscombe & P.T. Geach (eds.), (New York, 1971).

DORTER, K., 'First Philosophy: Metaphysics or Epistemology?', *Dialogue* 11, 1 (Mar. 1972).

DOVE, H.W.,

- contrib. to: *Monatsber. der Berl. Akad.* (July 29, 1841).

DOVE, H.W., (*cont.*)

- 'Ueber die Ursache des Glanzes und der Irridation, abgeleitet aus chromatischen Versuchen mit dem Stereoskop', *Pogg. Annalen*, 83, (1850).

DUMMETT, M., *Frege. Philosophy of Language*, (London, 1973).

DIJKSTERHUIS, E.J., *De Mechanisering van het Wereldbeeld*, (Amsterdam, 1950).

EINSTEIN, A., 'Zur Elektrodynamik bewegter Körper', *Ann. d. Physik*, 17 (1905).

ELKANA, Y., 'The Problem of Knowledge in Historical Perspective', *Proceedings of the 2nd International Humanistic Symposium*, (Athens, 1974), Hellenic Society for Humanistic Studies, International Centre for Humanistic Research, 2nd series, 24.

ERDMANN, B., 'Die philosophischen Grundlagen von Helmholtz' Wahrnehmungstheorie', in *Abhandlungen der Preussischen Akademie der Wissenschaften*, (Berlin, 1921).

FEIGL, H., 'Existential Hypotheses: Realistic versus Phenomenalistic Interpretations', *Phil. Sci.* 17 (1950), pp. 35-62.

FEYERABEND, P.K.,

- 'How to be a Good Empiricist—a Plea for Tolerance in Matters Epistemological', repr. in *The Philosophy of Science*, P.H. Nidditch (ed.), (Oxford University Press, 1968), pp. 12-39.
- 'Classical Empiricism', in *The Methodological Heritage of Newton*, R.E. Butts & J.W. Davis (eds.), (Oxford, 1970), pp. 150-70.

FICHTE, J.G., *Popular Works*, (London, Trübner, 1889).

FORTLAGE, K., 'Materialismus und Spiritualismus', *Blätter für literarische Unterhaltung* 30 (1856).

FRANKFURT, H.G., *Demons, Dreamers and Madmen: The Defense of Reason in Descartes' Meditations*, (Indianapolis, 1970).

GALENUS, C.,

- *Claudii Galeni opera omnia*, C.G. Kühn (ed.), 22 vols. (Leipzig, 1821-1833); (repr. Hildesheim, 1965). Vol. 5 (1823), pp. 181-805: *De placitis Hippocratis et Platonis..*
- *De placitis Hippocratis et Platonis*, Libri novem, Vol. I, Prolegomena critica, textum graecum, adnotationem criticum versionemque latinam continens, Iwan Mueller (ed.), (Leipzig, 1874).

GALILEO, G., *The Assayer (Il Saggiatore)*, in *Discoveries and Opinions of Galileo*, S. Drake (tr.), (Doubleday, 1957).

GEWIRTZ, A.,

- 'The Cartesian Circle', *Phil. Rev.* L, 4, (July 1941), pp. 368-95.
- 'The Cartesian Circle Reconsidered', *J. Phil.* 67, 19, (Oct. 8, 1970), pp. 668-85.

GRANT, E., (ed.), *A Source Book in Medieval Science*, (Cambridge, 1974).

GREGORY, F., *Scientific Materialism in Nineteenth Century Germany*, (Dordrecht, 1977).

GREGORY, R.L., *Concepts and Mechanisms of Perception*, (New York, 1974).

GROSSBERG, S., *Studies of Mind and Brain, Neural Principles of Learning, Perception, Development, Cognition, and Motor Control,* (Reidel, 1982).

GROSSETESTE, R., *Die philosophischen Werke des Robert Grosseteste, Bisschofs von Lincoln*, L. Baur (ed.), (Münster, 1912).

HACKER, P.M.S., 'Helmholtz's Theory of Perception: an Investigation into its Conceptual Framework', *forthcoming*.

HALL, T.S., 'Descartes' Physiological Method: Position, Principles, Examples', *Journal of the History of Biology* 3 (Spring 1970), pp. 53-79.

HAMM, J., *Das philosophische Weltbild von Helmholtz*, (Bielefeld, 1937), Inaugural Dissertation at Bonn.

HANSON, N.R., *Patterns of Discovery*, (Cambridge, 1969).

HEGEL, G.W.F., *Sämtliche Werke*, H. Glockner (ed.), 20 vols., (Stuttgart, 1928).

HELMHOLTZ, H. VON,

- 'Über Herrn D. Brewsters neue Analyse des Sonnenlichts', *Pogg. Annalen*, (1851), LXXXVI.
- *Handbuch der physiologischen Optik*, 3 vols., (Heidelberg, 1856/1860/1867), 875 pp.; 2nd ed., in 8 parts, (Hamburg, 1885-1894); 3d ed., 3 vols., (Hamburg, 1909-1911); Eng. tr. by J.P.C. Southall, *Helmholtz's Treatise on Physiological Optics*, 3 vols., from the 3rd German ed., published by the Optical Society of America (Rochester,. New York, 1924-25; Dover reprint, 3 vols. in 2, New York, 1962), 482, 479 and 746 pp.
- 'The Origin and Meaning of Geometric Axioms (II)', *Mind* 3 (1878), 10, pp. 212-25.
- *Wissenschaftliche Abhandlungen*, 3 vols., (Leipzig 1882, 1883 and 1895).
- 'Ueber die Theorie der zusammengesetzten Farben', (1852), in *Wissenschaftliche Abhandlungen*, H. von Helmholtz, (Leipzig, 1882, 1883 and 1895).
- *Vorträge und Reden*, 2 vols., (5th ed., Braunschweig, 1903).
- *Schriften zur Erkenntnistheorie*, ed. with extensive notes by P. Herz & M. Schlick (Berlin, 1921); tr. by M.F. Lowe as: *Epistemological Writings*, Boston Studies in the Philosophy of Science vol. XXXVII, R.S. Cohen & Y. Elkana (eds.), (Dordrecht & Boston, 1977).
- 'The Origin and Meaning of Geometric Axioms (I)', (1870), in *Selected Writings of Hermann von Helmholtz*, R. Kahl (ed.), (Middletown, Conn., 1971).
- 'Über das Sehen des Menschen', (1855), in *Philosophische Vorträge und Aufsätze*, H. Hörz & S. Wollgast (eds.), (Berlin, 1971).

HERING, E.,
- *Zur Lehre vom Lichtsinn* (1874).
- contrib. to *Sitzungsberichte der Wiener Akademie*, mathem.-naturw. Kl. LXIX, (1874).
- *Grundzüge einer Theorie des Lichtsinns*, (1874-1875).
- 'The Theory of Nerve Activity' (1898), in *On Memory, Lectures on the Specific Energies of the Nervous System* (Chicago, 1913[4]).

HERRNSTEIN, R.J. & E.G. BORING (eds.), *A Source Book in the History of Psychology*, (Cambridge, 1965).

HOLTON, G.,
- 'Einstein, Michelson, and the "Crucial" Experiment', *Isis* 60 (1969), pp. 133-97.
- *Thematic Origins of Scientific Thought: Kepler to Einstein*, (Cambridge, 1973).
- 'Mach, Einstein, and the Search for Reality', *Thematic Origins of Scientific Thought* (Cambridge, 1973), pp. 219-59.

HUME, D., *A Treatise of Human Nature*, L.A. Selby-Bigge (ed.), (Oxford, 1888).

ITZKOFF, S.W., *Ernst Cassirer: Scientific Knowledge and the Concept of Man*, (Notre Dame, 1975).

JULESZ, B., 'Texture and Visual Perception', *Scientific American*, (Febr., 1965).

KAHL, R., *Selected Writings of Hermann von Helmholtz*, (Middletown, Conn., 1971).

KANT, I., *Kritik der reinen Vernunft*, (Darmstadt, 1968).

KENNY, A.,
- *Descartes: A Study of His Philosophy*, (New York, 1968).
- 'The Cartesian Circle and the Eternal Truths', *J. Phil.* 67, 19, (Oct. 8, 1970), pp. 685-700.

KEPLER, J., *Gesammelte Werke*, herausgegeben im Auftrag der deutschen Forschungsgemeinschaft und der bayerischen Akademie der Wissenschaften unter der Leitung von W. von Dyck und M. Caspar, 18 Bde., (München, 1937-19...).

KLUCKHORN, P., *Die Idee des Menschen in der Goethezeit*, (Stuttgart, 1946).

KOENIGSBERGER, L., *Hermann von Helmholtz*, (Braunschweig, 1902-1903), 3 vols.; tr. into English by F.A. Welby, (Oxford, 1906).

KÖHLER, W., 'Die physische Gestalten', (1920), tr. in *A Source Book of Gestalt Psychology*, W.C. Ellis (ed.), (London, New York, 1938), pp. 17-54.

KOYRÉ, A., *Metaphysics and Measurement*, (London, 1968).

KOLERS, P.A., 'Subliminal Stimulation in Problem-Solving', *Amer. J. Psych.* 70 (1957), pp. 437-41.

KRAUSE, A., *Kant und Helmholtz: Über den Ursprung und die Bedeutung der Raumanschauung und der geometrischen Axiome*, (Lahr, 1878).

KRAUTHEIMER, R. & KRAUTHEIMER-HESS, T., *Lorenzo Ghiberti*, (Princeton, 1956).

KUHN, T.S.,

- 'Mathematical versus Experimental Traditions in the Development of Physical Science', *The Journal of Interdisciplinary History*, 7, (1976), pp. 1-31.
- *The Essential Tension*, (Chicago, 1977).

LACOSTA, L.M., 'La notion d'évidence et le sens commun: Fénelon et Reid', *Journal of the History of Philosophy* 15, 3 (July 1977), pp. 293-307.

LAKATOS, I.,

- 'Falsification and the Methodology of Scientific Research Programmes', in *Criticism and the Growth of Knowledge*, I. Lakatos & A. Musgrave (eds.), (Cambridge, 1970), pp. 91-135.
- 'History of Science and its Rational Reconstructions', in *P.S.A. 1970,*, *Boston Studies in the Philosophy of Science* 8, R.C. Buck & R.S. Cohen (eds.), (Dordrecht, 1971), pp. 91-196.

LAMETTRIE, J.O. DE, *Histoire naturelle de l'âme* (also known as *Traité de l'âme*), (The Hague, 1745).

LAND, E.H.,

- 'The Retinex', *Am. Scientist*, vol. 52 (1964), pp. 247-64.
- 'The Retinex Theory of Color Vision', *Scientific American*, vol. 237 (1977), 6, pp. 108-28.

LAND, J.L., 'Kant's Space and Modern Mathematics', *Mind* 2 (1877), pp. 38-46.

LANGE, F.A., *The History of Materialism*, 1st German ed. 1865; tr. by E.C. Thomas with an introduction by B. Russell, (London, 1925).

LEJEUNE, A.,

- 'Les tables de réfraction de Ptolémée', *Annales de la société scientifique de Bruxelles*, ser. 1, vol. 60 (1946), pp. 95 ff.
- *Euclide et Ptolémée. Deux stades de l'optique géométrique grecque*, Université de Louvain, Recueil de travaux d'histoire et de philologie, ser. 3, fasc. 31, (Louvain, 1948).
- 'Les recherches de Ptolémée sur la vision binoculaire', *Janus* 47 (1958), pp. 79-86.

LEMAY, R., *Abu Ma'shar and Latin Aristotelianism in the Twelfth Century*, (Beirut, 1962).

LENIN, W.I., *Materialismus und Empirio-kritizismus*, in *Werke*, Bd. 14, (Berlin, 1962); tr. as *Materialism and Empirio-criticism*, A. Fineberg (tr.), (Moscow, 1962).

LENZEN, V., 'Helmholtz's Theory of Knowledge', in *Studies and Essays offered to George Sarton*, M.F. Ashley Montagu (ed.), (New York, 1947), pp. 301-19.

LEWIS, C.I.,
- *Mind and the World Order*, (New York, 1929).
- *An Analysis of Knowledge and Valuation*, (La Salle, 1946).

LINDBERG, D.C.,
- 'Alkindi's Critique of Euclid's Theory of Vision', *Isis*, 62 (1971).
- *Theories of Vision from Al-Kindi to Kepler*, (Chicago, 1976).

LINDBERG, D.C. & STENECK, N.H., 'The Sense of Vision and the Origins of Modern Science', in *Science, Medicine and Society in the Renaissance, Essays to honor Walter Pagel*, A.G. Debus (ed.), (New York, 1972).

LOCKE, J.,
- *Examination of the Opinions of P. Malebranche*, in *Works*, (London, 1823).
- *An Essay concerning Human Understanding*, M. Cranston (ed.), (London, 1965).

LOTZE, H., *Grundzüge der Psychologie*, (Leipzig, 1882).

LUCE, A.A., *Berkeley and Malebranche, A Study in the Origins of Berkeley's Thought*, (Oxford, 1967).

LUCRETIUS, T.C., *De rerum natura*, R.E. Latham (tr.), (Penguin Books, 1951).

MAGENDIE, F.,
- 'Expérience sur les fonctions des racines des nerfs rachidiens', *Journal de physiologie expérimentale et pathologique* 2 (1822), pp. 276-9.
- 'Expériences sur les fonctions des racines des nerfs qui naissent de la moëlle épinière', *Journal de physiologie expérimentale et pathologique* 2 (1822), pp. 366-71.

MALEBRANCHE, N.,
- *Oeuvres complètes de Malebranche*, publié par MM. de Genoude et de Lourdoueix, (Paris, 1837).
- *Oeuvres de Malebranche*, P. Costabel & A. Robinet (eds.), vol. XVII, (Paris, 1960); G. Rodis-Lewis (ed.), vol. III, (Paris, 1964).

MANDELBAUM, M.,
- 'Locke's Realism', *Philosophy, Science, and Sense Perception*, (Baltimore, 1964), pp. 1-60.
- *History, Man and Reason, A Study in Nineteenth-Century Thought*, (Baltimore, 1971).

MAULL, N.L., 'Cartesian Optics and the Geometrization of Nature', *The Review of Metaphysics*, 32, 2 (Dec. 1978).

MAUROLYCUS, F., *The Photismi de Lumine of Maurolycus: A Chapter in Late Medieval Optics*, H. Crew (tr.), (New York, 1940).

MERSENNE, M., *Harmonie universelle*, (Paris, 1636-1637).
MERZ, J.T., *A History of European Thought in the Nineteenth Century*, 4 vols., (1904-1912), (Dover reprint: New York, 1965).
MILL, J., *Analysis of the Phenomena of the Human Mind*, 2 vols., I, (London, 1829).
MILL, J.S.,
• *An Examination of Sir William Hamilton's Philosophy*, (London, 6th ed., 1889).
• *A System of Logic Ratiocinative and Inductive. Being a Connected View of the Principles of Evidence and the Methods of Scientific Investigation.*, J.M. Robson (ed.), 2 vols., (Toronto University Press, 1973-1974).
MOLESCHOTT, J., *Ursache und Wirkung in der Lehre vom Leben*, (Giessen, 1867).
MOLYNEUX, W. *Dioptrica Nova: A Treatise of Dioptrics*, (London, 1692).
MONTAIGNE, M. DE, *The Essays of Montaigne*, J. Florio (tr.), (The Modern Library, New York).
MÜLLER, J., *Handbuch der Physiologie des Menschen*, (Coblenz, 1838); tr. by W. Baly as *Elements of Physiology*, (London, 1842).
NEWTON, I., 'Reply to Lucas', in *Isaac Newton's Papers and Letters on Natural Philosophy*, I.B. Cohen (ed.), (Cambridge, 1958).
NIETZSCHE, F., *The Will to Power*, W. Kaufmann (ed.), (New York, 1968).
PANUM, P.L.,
• *Physiologische Untersuchungen über das Sehen mit zwei Augen*, (Kiel, 1858).
• 'Ueber die einheitliche Verschmelzung verschiedenartiger Netzhauteindrücke beim Sehen mit zwei Augen', *Reicherts Arch. für Anat. u. Physiol.* (1861), pp. 63-111, 178-227.
PIAGET, J., *Psychology and Epistemology, towards a Theory of Knowledge*, A. Rosin (tr.), (Penguin Books, 1977).
PLOTINUS, *The Enneads*, S. Mackenna (tr.), revised ed. by B.S. Page, (London, 1962^3).
POINCARÉ, H., *The Value of Science*, G.B. Halsted (tr.), (New York, 1907).
POPPER, K., *Conjectures and Refutations*, (London, 1963).
PRICE, H.H., *Perception*, (London, 1932).
PROCLUS, *A Commentary on the First Book of Euclid's Elements*, I, 40, G.R. Morrow (tr.), (Princeton, 1970).
QUINE, W.V.O.,
• 'On What There Is', in id., *From a Logical Point of View*, (New York, 1953), pp. 1-19.
• 'The Nature of Natural Knowledge', in *Mind and Language*, S. Guttenplan (ed.), (Oxford, 1975), pp. 67-82.
RANDALL JR., J.H., *The Making of the Modern Mind*, (Boston, 1926).

RAAB, D.H., 'Backward Masking', *Psychological Bulletin* 60 (1963), 2, pp. 118-29.
REGIS, P.S., *Système de philosophie*, 3 vols., (Paris, 1690).
REICHENBACH, H., *Experience and Prediction*, (Chicago, 1938, 1947).
RIEHL, A.,
• 'Helmholtz in seinem Verhältnis zu Kant', *Kant-Studien* 9 (1904), pp. 261-285.
• *Führende Denker und Forscher*, (Leipzig, 1924).
ROCK, I.,
• 'In Defense of Unconscious Inference', in W. Epstein (ed.), *Stability and Constancy in Visual Perception: Mechanisms and Processes*, (New York, 1977).
• *The Logic of Perception*, (Cambridge, 1983).
RONCHI, V., *Optics, The Science of Vision*, (New York University Press, 1957).
ROSEN, E., 'The Invention of Eyeglasses', *Journal of the History of Medicine and Allied Sciences* XI (1956), pp. 13-53, 183-218.
RUMELHART, D.E. & MCCLELLAND, J.L. (eds.), *Parallel Distributed Processing, Explorations in the Microstructure of Cognition*, 2 vols., (MIT Press, 1986).
RUSSELL, B.A.W.,
• *Our Knowledge of the External World*, (London, 1914).
• *Inquiry into Meaning and Truth*, (London, 1940).
• *Human Knowledge: Its Scope and Limits*, (London, 1948).
SABRA, A.I., 'Sensation and Inference in Alhazen's Theory of Visual Perception', in *Studies in Perception*, P.K. Machamer & R.G. Turnbull (eds.), (Columbus, 1978).
SAMBURSKY, S., *Physics of the Stoics*, (London, 1959).
SCHIFFER, S.R., 'Descartes on his Essence', *Phil. Rev.* 85, 1 (Jan. 1976).
SCHLEIDEN, M.,
• *Über Schellings und Hegels Verhältnis zur Naturwissenschaft* (1844).
• *Über den Materialismus der neueren deutschen Naturwissenschaft, sein Wesen und seine Geschichte*, (Leipzig, 1863).
SCHLICK, M., *Allgemeine Erkenntnislehre*, (Berlin, 1918).
SCHRAMM, M., 'Zur Entwicklung der physiologischen Optik in der Arabischen Literatur, *Sudhoffs Archiv für Geschichte der Medizin und der Naturwissenschaften*, 43, 4 (1959).
SCHWERTSCHLAGER, J., *Kant und Helmholtz erkenntnistheoretisch verglichen*, (Freiburg-im-Breisgau, 1883).
SHAPERE, D., *Galileo: A Philosophical Study*, (Chicago, 1974).
SLUGA, H.D.,
• 'Frege and the Rise of Analytic Philosophy', *Inquiry*, 18 (1975), pp. 471-87.

• *Gottlob Frege*, (London, 1980).
SMITH, G., 'Visual Perception: An Event over Time', *Psychol. Review* 64, 5 (1957), pp. 306–13.
SMITH, N.K., *New Studies in the Philosophy of Descartes*, (London, 1952).
SOLMSEN, F., 'Greek Philosophy and the Discovery of the Nerves', *Museum Helveticum* 18 (1961), pp. 150-67; 169-97.
STRATTON, G.M., 'Vision without Inversion of the Retinal Image', *Psych. Rev.* 4, (1897), pp. 341-481.
STRONG, D.S., *Leonardo da Vinci on the Eye*,The Ms D... translated into English and annotated with a study of Leonardo's theories of optics, D.S. Strong, (Ph.D. diss., UCLA, 1967).
STROUD, B., *Hume*, (London, 1977).
TEMKIN, O., 'Materialism in French and German Physiology of the Early Nineteenth Century', *Bull. Hist. Medicine* 20 (1946), pp. 322-7.
THORNDIKE, L., *A History of Magic and Experimental Science*, 8 vols., (New York, London, 1923-1956).
TREITSCHKE, H., *History of Germany in the Nineteenth Century*, E. & C. Paul (trs.), 7 vols., (London, 1919).
TURNER, R.S., 'Hermann von Helmholtz and the Empiricist Vision', *J. Hist. Behavioral Sci.*, 13 (1977).
VESCOVINI, G.F., *Studi sulla prospettiva medievale*, (Turin, 1965).
VINCI, LEONARDO DA,
• *Literary Works*, J.P. Richter & I.A. Richter (eds./trs.), (Oxford, 1939).
• *Selections from the Notebooks of Leonardo da Vinci*, I.A. Richter (ed./tr.), (Oxford, 1952).
• *Of the Eye*, in Strong (1967); also in 'Leonardo da Vinci: Of the Eye', N. Ferrero (tr.), *Am. J. Ophthalm.*, ser. 3, 35 (1952).
VIRCHOW, R., 'Die Gründung der Berliner Universität und der Übergang aus dem philosophischen in das naturwissenschaftliche Zeitalter', in *Idee und Wirklichkeit einer Universität. Dokumente zur Geschichte der Friedrich-Wilhelms-Universität zu Berlin,* Wilhelm Weischedel (ed.), 2 vols., (Berlin, 1960).
WEBB, T.E., *The Intellectualism of Locke*, (Dublin, 1857).
WHITEHEAD, A.N., *Science and the Modern World*, (New York, 1925).
WHYTE, L.H., *The Unconscious before Freud*, (New York, 1960).
WILLEY, T. E., *Back to Kant*, (Detroit, 1978).
WILSON, E.O., *Sociobiology, The New Synthesis*, (Cambridge, Mass., 1975).
WILSON, M.D., *Descartes*, (London, 1978).
WITELO, *Perspectiva*, in *Opticae thesaurus Alhazeni Arabis libri septem... Item Vitellonis Thuringopoloni libri decem*, (Risner: Basel, 1572; repr. New York, 1972).
WITTGENSTEIN, L., *Tractatus Logico-Philosophicus*, (London, 1922).

WOLF, R., *Geschichte der Astronomie*, (München, 1877).
WUNDT, R., *Grundriss der Psychologie*, (Leipzig, 1905).
YOUNG, T., 'On the Theory of Light and Colours', *Philosophical Transactions of the Royal Society of London*, (1802), 92, pp. 18-21.
ZIEGLER, T., *Die geistigen und sozialen Strömungen des neunzehnten Jahrhunderts*, (Berlin, 1911).
ZILSEL, E., 'The Origins of William Gilbert's Scientific Method', *Journal of the History of Ideas* 2 (1941), pp. 1-32.

LIST OF ABBREVIATIONS

AG — *Descartes, Philosophical Writings*, E. Anscombe & P.T. Geach (eds.), (New York, 1971).

ASP — Alhazen, *De aspectibus*, in *Opticae thesaurus. Alhazeni Arabis libri septem...*, (Risner: Basel, 1572; repr. New York, 1972).

AT — *Oeuvres de Descartes*, C. Adam & P. Tannery (eds.), 12 vols., (Paris, 1897-1910).

EW — SE, tr. by M.F. Lowe as *Epistemological Writings*, R.S. Cohen & Y. Elkana (eds.), Boston Studies in the Philosophy of Science, vol. XXXVII, (Dordrecht & Boston, 1977).

HB — *A Source Book in the History of Psychology*, R.J. Herrnstein & E.G. Boring (eds.), (Cambridge, 1965).

HR — *The Philosophical Works of Descartes*, 2 vols., E.D. Haldane & G.R.T. Ross (eds.), (New York, 1973).

Kahl — R. Kahl, *Selected Writings of Hermann von Helmholtz*, (Middletown, Conn., 1971).

OM — *Oeuvres de Malebranche*, vol. XVII, (Paris, 1960), P. Costabel, A. Cuvillier & A. Robinet (eds.); vol. III, (Paris, 1964), G. Rodis-Lewis (ed.)

PL — F. Maurolycus, *The Photismi de Lumine of Maurolycus: A Chapter in Late Medieval Optics*, H. Crew (tr.), (New York, 1940).

PO — *Helmholtz's Treatise on Physiological Optics*, 3 vols., tr. into English by J.P.C. Southall, from the 3d German edition of Hermann von Helmholtz, *Handbuch der physiologischen Optik*, 3rd ed., in 3 vols. (Hamburg, 1909-1911); published by the Optical Society of America, (Rochester/New York, 1924-1925); Dover reprint: 3 vols. in 2, New York, 1962). PO refers to the Southall edition; all other editions are referred to as PO followed by the corresponding year of publication.

RV — *Recherche de la vérité*, in *Oeuvres complètes de Malebranche*, publié par MM. de Genoude et de Lourdoueix, Tôme Premier, (Paris, 1837).

SE — Hermann von Helmholtz, *Schriften zur Erkenntnistheorie*, ed. with extensive notes by P. Herz & M. Schlick, (Berlin, 1921).

T W — H. von Helmholtz, 'Die Tatsachen in der Wahrnehmung', in SE.

VR — Hermann von Helmholtz, '*Vorträge und Reden*, 2 vols., (5th ed., Braunschweig, 1903).

W A — H. von Helmholtz, *Wissenschaftliche Abhandlungen*, 3 vols., (Leipzig, 1882, 1883, 1895).

ARISTOTLE — All references to works by Aristotle are taken from *The Student's Oxford Aristotle*, W.D. Ross (ed.), (New York, 1942), unless otherwise indicated.

CLAUDIUS GALENUS — References to his *De placitis Hippocratis et Platonis* are taken from *C. Galenus, De placitis Hippocratis et Platonis*, libri novem, I. Mueller (ed.), Vol. I, Prolegomena critica, textum graecum, adnotationem criticam versionemque latinam continens, (Leipzig, 1874). References to the Mueller edition will be followed, in parentheses, by the corresponding reference in the Kühn edition of Galen's works.

G.W.F. HEGEL — References to Hegel's works refer to his *Sämtliche Werke*, 20 vols., H. Glockner (ed.), (Stuttgart, 1928), unless otherwise indicated.

INDEX